A PRACTICAL GUIDE TO INSPECTING

ROOFS

By Roy Newcomer

CONTENTS

INTRODUCTION

My background includes many years in construction and several more as the owner of a Century 21 real estate franchise. In 1989, I started a home inspection company that grew larger than I ever imagined it could. Training my own staff of inspectors to the highest inspection standards led to my teaching home inspection seminars across the country and developing study courses, books, and videos for home inspectors. The American Home Inspectors Training Institute was founded as a result of my desire to share this experience and knowledge in home inspection.

The *Practical Guide to Inspecting* series is intended for both beginning and experienced home inspectors. So if you're studying home inspection for the first time or are using the materials as a refresher, these guides should be of assistance to you.

I've created these guides to include all aspects of home inspection. Not only a broad technical background in home systems, but the other things you need to know in order to perform a *good* inspection of those systems. They lay out technical information, guidelines for the inspection, how-to instructions for inspecting system components, and the defects, deficiencies, and problems you'll be looking for during the inspection. I've also included some advice on how to report your findings to the home inspection customer.

I've been a member of several professional organizations for a number of years, including ASHI® (American Society of Home Inspectors), NAHI™ (National Association of Home Inspectors), and CREIA® (California Real Estate Inspection Association). I am a great supporter of those organizations' quest to promote excellence in home inspection.

I encourage you to follow the standards of the organization to which you might belong, or any state regulation that might take precedent over the standards used here. Use the standards in this book as a general guide for study and apply the standard or state regulation that applies to you.

The inspection guidelines presented in the Practical Guides are an attempt to meet or exceed standards and regulations as they exist at the revision date of the guides.

There's a lot to learn about home inspection. For beginning inspectors, there are some *hands-on exercises* in this guide that should be done. I'm a great believer in learning by doing, and I hope you'll try them. There are also some of my *personal inspection stories* to let you know what it's really like out there.

The *inspection photos* referenced in this text can also be found on www.ahit.com/photos. You'll read the story about each one as you go along. Be sure to watch for my *Don't Ever Miss* lists. I've included them to alert home inspectors to report those defects (if found during the inspection) in the inspection report. If missed, these items are often the cause for lawsuits later. Finally, to help you see how you're doing as you study this guide, I've included some *worksheets*. The answers are given for each one for self checking. Give them a try. Checking yourself can help you lock important information in your mind. There's also a *final exam* that you can complete and send in to us. Many organizations and states have approved this book for continuing education credits. Submit the exam with the required fee if you need these credits.

In total, the *Practial Guide to Inspecting* series covers all aspects of the general home inspection. Each guide covers a major aspect of the inspection, as their titles show:

Electrical
Exteriors
Heating and Cooling
Interiors, Insulation, Ventilation
Plumbing
Roofs
Structure

If you are interested in other titles in the series, please call us at the American Home Inspectors Training Institute to order them. Call toll free at 1-800-441-9411.

Roy Newcomer

INSPECTING ROOFS

Chapter One

THE ROOF INSPECTION

While the structure of the house ranks number one in importance to the home inspector when judging how sound a home is, a wet basement and a poor roof tie for first place in customer concerns. The customer will concentrate on the roof covering, wanting to know if it has to be replaced. The home inspector inspects the roof for the condition of the covering, soundness of structure, water penetration, condition of the chimney and flashings, and effectiveness of drainage.

Inspection Guidelines and Overview

These are the standards of practice that guide the inspection of the roofing system. Please read them carefully.

Roofing System	
OBJECTIVE	To identify major deficiencies in the condition of the roofing system.
OBSERVATION	<u>Required to inspect and report:</u> • Roof coverings and type • Roof drainage systems • Flashings • Skylights, chimneys, and roof penetrations • Signs of leaks or abnormal condensation on building components
ACTION	<u>Not required to observe:</u> • Attached accessories including, but not limited to, solar systems, antennae, and lightning arrestors. <u>Required to:</u> • Report the methods used to observe the roofing. <u>Not required to:</u> • Walk on the roofing when walking could damage the property or be unsafe to the inspector.

Pages 1 to 8 outline the content and scope of the roof inspection. It's an overview of the inspection, including what to observe, what to describe, and what specific actions to take during the inspection. Study these guidelines carefully. These pages also present special cautions about inspecting the roof. Please read them and take heed of the potential of injuring yourself or damaging the roof during the roof inspection.

INSPECTING THE ROOF

- Roof coverings
- Roof drainage systems
- Flashings
- Chimneys
- Roof penetrations
- Water penetration and condensation

Guide Note

This guide presents the exterior roof inspection, not the inspection of the roof's structure. That's covered in another of our guides — A Practical Guide to Inspecting Structure.

Not every single detail on what to inspect and what to report is given in these guidelines. But they do serve as a good outline of what is expected from the home inspector during the inspection of the roof.

The **main purpose** of the roof is to protect the house from the elements — wind, snow, rain, and the sun. The home inspector must determine how well the roof can do its job. The roof also provides *some* protection from falling objects, although not enough to protect the house from falling tree limbs. Contrary to popular belief, the roof does *not* serve as an insulator to keep out the cold.

This guide presents information on the **exterior roof inspection**. This inspection goes hand in hand with the inspection of the roof's structure and underlayment, which was presented in another of our guides — *A Practical Guide to Inspecting Structure*. The home inspector cannot really separate the structural inspection from the exterior inspection of the roof. When you're on the roof, you need to be conscious of what's going on underneath. When you're inside the attic, you need to remember the potential danger signs you saw on the outside.

Here is an overview of the exterior roofing inspection as stated in the guidelines:

- **Access:** Probably the most important information the home inspector can record in his or her inspection report is to say how much of the roof was visible and from where the roof was inspected. In fact, the guidelines state that the home inspector is **required to report methods used to observe the roof**. This is for the home inspector's protection from lawsuits filed later. New homeowners with a leaking roof want to blame someone, and that blame almost always falls on the home inspector. If the report clearly states that only 30% of the roof was visible due to snow coverage and that the roof had to be inspected from the ladder at the roof's edge, then the inspector is legally covered.

  We'll talk much more about when it's safe to get on a roof and when it's not. Note that the guidelines say that the home inspector is **not required to walk on the roof** when walking could damage the property or be unsafe.

- **Roof coverings:** The home inspector examines the roof of the house and garage to determine its approximate age and identifies its style — gable, mansard, flat, and so on. The **type of roof covering** is described. Examples are asphalt shingles, wood shingles and shakes, roll roofing, built-up roofing, metal, and tile. The home inspector inspects the **condition** of the roofing, watching out for defects such as improper installation curling, cracking, cupping, missing shingles, moss buildup, nail pops, and water ponding.

Customers also expect the home inspector to estimate the roof covering's remaining **useful lifetime**. We use a simple system of reporting expected lifetime in our own inspection report. A report of **satisfactory** means the roof covering will last more than 5 years, **marginal** is used to indicate that the roof covering will have to be replaced within the next 5 years, and **poor** is used when we determine the roof covering needs to be replaced now or very soon. (That's explained to the customer in the report.)

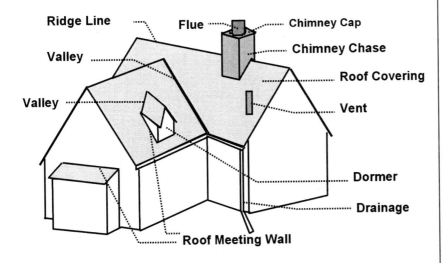

- **Flashings:** The home inspector examines the flashings used in the roofing system. Flashings are necessary to protect the roof in **valleys** where different roof sections meet, where roof and walls meet at a lower roof or a **dormer**, around chimneys and other penetrations through the roof, and at the edge of the roof. The inspector is required to identify the **materials** used such as copper, aluminum, or galvanized steel. The inspector must report the **condition** of flashings, whether the flashing is rusted, bent, loose, leaking, improperly installed, or missing.

Definitions

<u>Flashings</u> are a type of sheet metal or plastic used at joints between building components to prevent water penetration. Flashings in the roofing system are used in valleys, at hips and ridges, where roof and walls meet, and around chimneys and penetrations through the roof such as vents.

The <u>ridge</u> is the horizontal intersection of two sloping roof surfaces. A <u>valley</u> is the trough formed by the junction of two sloping sides of the roof.

A <u>dormer</u> is a structure built out from the roof slope, having its own roof and walls.

The chimney chase is the outer construction that encloses the flue. The chimney flue is the channel that carries gases, fumes, and smoke from furnaces and fireplaces. The chimney cap is a concrete or metal covering for the top of the chase.

Where flashings are covered with tar, they should be reported as not visible for inspection.

- **Chimneys:** The home inspector inspects the chimneys, noting their location on the roof. It is a good idea to record from where the chimneys were viewed for the home inspector's protection. The **chase material** is identified and its condition noted. The inspector checks for cracks, deteriorating mortar, and leaning. The **chimney cap** over the chase is inspected to see if it is rusted, cracked, broken, or missing. The **flue material** is identified and the flue checked for condition. The home inspector may recommend that the flue should be cleaned and evaluated. Of course, chimney flashings are inspected.

- **Roof penetrations:** Whenever a vent or skylight is penetrating through the roof, the home inspector examines the flashings around the penetration for water tightness. Skylights are inspected for condition. The home inspector may note cracked or broken glazing or condensation between the lights.

NOTE: The guidelines state that the home inspector is **not required to observe attached accessories** such as antennae and solar collectors. But if these accessories have a detrimental effect on the roof, it should be reported. The home inspector examines the attachments of such accessories to the roof surface to see if they are securely attached, have proper flashings, or are the cause of leaks. Some homeowners have an antenna strapped to the chimney or a vent, which can result in damage to the chimney or vent. Where heavy accessories are present, the home inspector checks the attic to see if the roof's structural members can support this additional weight.

- **Water penetration:** Leaks through the roof are not always discovered from the exterior. The home inspector may have to enter the attic to be able to see what damage has been done. But from the surface, the home inspector watches out for signs of leakage in the roof coverings and in areas where flashings are present.

- **Drainage system:** Water should not only drain off the roof, it must drain in such a way that it doesn't pool around the foundation and cause water penetration into the basement. Gutters and downspouts are inspected carefully to be sure they're not rusted or rotted, disconnected, or full of debris. The home inspector also checks extensions to downspouts and splash blocks to make sure water isn't washing against the foundation.

- **Structure:** Although the roof's structure is examined from the attic, the home inspector watches for any signs of structural problems while examining the roof from the outside. A wavy surface to the roof can indicate rafters were improperly installed or deteriorating roof sheathing. Soffits that pull away from the wall are an indication of rafter spread. A sagging or broken ridge line can be a sign of other structural problems. Falling through the roof is definitely a sign of deterioration of some of the roof's structural members!

Inspection Equipment

When inspecting the roof from the exterior, the home inspector needs the following:

A ladder to get on the roof
Binoculars for inspecting the roof from below
A probe such as a screwdriver
Soft, rubber soled shoes or boots that will grip the roof surface

A home inspector travels fairly light, without a great number of tools and without tools that take up too much space. The ladder is the only exception. Most home inspectors use their own vehicles to get to the inspection site. You must have a long enough ladder, but you don't want that to dictate what sort of vehicle you drive.

We recommend a compact **extension ladder** called the Little Giant®, which can also be used as a step ladder, perfect for stepping up on when inspecting soffits and fascia boards around the house. For getting on the roof, you'll need at least 17' of ladder, which the Little Giant® gives you. No matter what brand

---

INSPECTION TOOLS
- Solid and safe ladder
- Binoculars
- Probe
- Proper shoes

Personal Note

"One of my inspectors was inspecting a property on a frosty morning. He got up on the roof from the south side where the frost had melted off. As he was inspecting the chimney, he stepped over the ridge to the north side without paying attention to the roof surface.
"You guessed it. The north side was still frost covered and terribly slick. The inspector slid down the roof and rolled off into a snow bank. Unhurt, thank goodness.
"The inspector walked around to the south side again, where the ladder was. The Realtor smiled and asked, 'Did you get off the roof the way I think you got off?' We still tease him about it."

Roy Newcomer

of ladder you buy, make sure it fits safely in your vehicle, gives you the necessary extension, and is a good solid and safe one.

Inspection Concerns

The best home inspection is conducted when the customer is present throughout the inspection, following the home inspector through each step of the inspection. In fact, we consider having the customer present to be the #1 rule of home inspection. Communication after the inspection is over is never as good as talking to the customer during the inspection. The educational process works best when the customer is present to see what is being inspected and what the findings are.

There is one obvious exception to that rule that should be stated up front. *Never, never, never* allow the customer to get on your ladder or get up on the roof. You must consider yourself **responsible for your customer's safety**. There are too many horror stories that home inspectors tell when they get together. Inspectors fall off roofs. Some fall through roofs into the attic. Inspecting the roof is the most dangerous part of the inspection.

There are two other concerns the home inspector should keep in mind when inspecting the roof:

- The inspector's own safety
- Damage to the roof

Some standards of practice explicitly state that the inspector is *not required to walk on the roofing when walking could damage the property or be unsafe to the inspector.* Please take heed of these instructions. You don't want to hurt yourself or do any damage to the roof. The following list briefly discusses some instances where the home inspector should exercise caution. Some are regarding the inspector's safety, while others are listed here because walking on the roof can damage the roof covering. (We'll talk about specific examples of roof coverings again in pages 18 to 43.)

- **Too steep:** Some roofs are just too steep to walk on safely. You'll have to decide for yourself what you can safely navigate. Remember, walking *up* a steep roof is easier than walking *down* a steep roof. Make a decision about what's too steep before you get up on the roof. That's better than getting into trouble when you want to get off.

- **Snow covered:** Do not walk on a snow-covered roof. They can be very slippery, and home inspectors have been known to fall off. And snow covers up defects the inspector can't see. A heavy **frost** on a roof can also pose a dangerous, slippery roof surface.

- **Wet wood shingles or shakes:** A wood roof becomes slippery in the rain. Even after a shower, when the shingles or shakes are thoroughly saturated with water, they're still slippery. The home inspector should exercise caution even when walking on a dry wood roof. Usually there's no problem, but even some dry roofs can be slippery. The home inspector should not step onto deteriorated wood shingles and shakes because of the potential of breaking them.

- **Moss covered:** Moss creates a slippery surface. You'll often find mossy roofs on the north side of a house or in areas under overhanging trees. Don't get up on a roof that is entirely covered with moss. If there are small patches of moss, be careful not to step on them.

- **Tile roof:** A tile roof is typically not meant to be walked on. You simply can't walk on clay tiles without damaging them. Inspect a tile roof from the ladder at several places around the house.

- **Slate roof:** Never walk on a slate roof during the inspection because of the potential to damage it. Slate is a brittle roof covering that can crack and break if you walk on it.

- **Asbestos-cement shingles:** This roof covering is similar to slate in its brittle quality. Don't walk on this type of roof either. You can damage it and you won't find replacement shingles.

- **Curled asphalt shingles:** When asphalt shingles curl under at their tabs, it's a sign that the roofing material has reached the end of its useful lifetime. Don't get up on such a roof. You already know the condition is poor. Shingles in this condition will only crack and break under your step. Better to be able to point out the condition than to explain to the customer why you've destroyed the shingles.

DON'T GET ON THE ROOF

- Too steep
- Covered with snow
- Wet wood shingles or shakes
- Covered with moss
- Tile roof
- Slate roof
- Asbestos-cement shingles
- Curled asphalt shingles
- Corrugated plastic
- Soft or spongy
- And be cautious over cathedral ceilings!

Personal Note

"One of my inspectors walked up a steep roof with no problem. When it was time to get down, he began sliding and grabbed the chimney to stop himself. There was no way he could get off the roof without sliding right off.

"A neighbor finally noticed him and threw him a hose to wrap around the chimney enabling him to work his way down safely. But not before the real estate agent had snapped his picture! He made the broker's monthly newsletter — 'Home Inspector Stranded on Roof.' Of course, when he visits this broker, they don't let him forget it."

Roy Newcomer

Different types of roof covering are defined and discussed in pages 18 to 43. Warnings about whether or not to walk on a particular type of roof are repeated in these pages.

__Photo #1__ shows an example of a __roof you should not walk on__. First of all, this roof has a steep pitch to it. Second, the roof has quite a bit of snow cover. And last, this is a slate roof that shouldn't be walked on in any case. How do you inspect such a roof? Inspecting from the ground with binoculars should always be your last choice. Inspecting the roof from a ladder at the roof edge is much better. However, the bushes in front of the house were in the way. We really struggled to get the ladder behind the bushes at the side of the chimney to get a look at that valley as close as possible. Then we were able to squeeze the ladder in to get a close look at the valley on the other side of the gable. The rest of the roof was inspected from the ladder at several other positions around the house.

- **Corrugated plastic:** Some homes have a translucent corrugated plastic roof over patios or carports. This material is not meant to support the weight of someone walking on it. Don't walk on this type of roof.

- **Soft and spongy:** If you're ever on a roof and the surface feels soft or spongy to you, get off the roof immediately. Go to the attic and investigate the situation from below. Only if you confirm a solid roof structure from the attic should you consider getting back up on the roof.

- **Cathedral or Vaulted ceilings:** Use special caution on a roof over a cathedral ceiling. Often they're improperly insulated and ventilation is so bad under the roof that the plywood sheathing delaminates and is considerably weakened. We've had the experience of our home inspection staff stepping boldly around a roof over a cathedral ceiling and putting a foot right through it. It's better to test each step before you put your full weight down. Walking close to the ridge (but not on it) and close to the valleys (but not in them) will help.

#1 Roof you should not walk on

It should be noted with **Photo #1** that we made sure to report only 25% visibility of this roof (the back side of the roof had even more snow). We also recorded viewing the roof from the ladder at the edge of the roof.

Chapter Two

GENERAL INFORMATION

The home inspector should be familiar with different types of roof styles and some of the terms used when referring to roofs.

Terms

The **slope** of a roof is also called the pitch. The slope is defined as the ratio of the **rise** of the roof to the **run** of the roof. The rise is the vertical height of the roof; the run is the horizontal length to the center of the roof (1/2 the whole span).

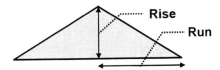

SLOPE = $\dfrac{\text{RISE}}{\text{RUN}}$

The convention is to report the ratio of rise to run over 12' of run. For example, a roof that rises 2' over the run of 12' would be called a slope of 2/12, which is said "two in twelve."

2/12 Slope 2' 12'

3/12 Slope 3' 12'

Roofs that have a slope of less than 2/12 are considered to be **flat roofs**. Flat roofs, of course, should not be perfectly flat. They should have a slight pitch for drainage purposes. Roofs with a slope between 2/12 and 4/12 are considered to have a **low slope**. What would be considered **conventional** roofing systems are those with a slope of 4/12 or greater.

4/12 Slope 4' 12'

6/12 Slope 6' 12'

Various roof coverings are not appropriate for certain slopes. Wood shingles should not be used on roofs with a slope less than 4/12. A greater slope can add years to the life of some coverings, that is, the steeper it is the longer the roofing will last. Again, wood shingles may last 20 years on a 4/12 roof and 30 years on a 6/12 roof. That's because water moves off a steeper roof faster, and the lower sloped roof is more likely to leak. The flat roof, for which no kind of shingles are appropriate, must use a sealed roofing material such as built-up or roll roofing.

Guide Note

Pages 9 to 14 present some roofing terms and types of roofs.

Definitions

The <u>slope</u> of a roof is the ratio of its rise to its run, where the <u>rise</u> is the vertical height of the roof and the <u>run</u> is the horizontal length of the roof to the center point. The slope is normally expressed with the measure of the rise over a run of 12'. A slope of 3/12 is said "three in twelve."

The term <u>square</u> is used to indicate the amount of roofing material used to cover 100 square feet of roof surface.

SOME ROOF STYLES

- Gable
- Gambrel
- Hip
- Mansard
- Shed
- Flat

For Beginning Inspectors

It's time to get in the car and take a ride around town looking at roofs. Keep an eye out for interesting variations of the basic styles presented on this page. You may find the saltbox which is a gable roof with one side having a longer run than the other. Or you might see a hip roof with a gabled section along the ridge line.

Roofers use the term **square** as a shortcut to mean the amount of roofing material used to cover 100 square feet of roof's surface. They may talk about so many squares needed to cover the roof. Builders will also think in terms of how much a square of a particular roof covering weighs when determining how much weight the roof structure must support. This becomes important when additional layers of roofing are added over the original. A roof built to support so many pounds per square may not be able to support a second or third layer of roofing.

Types of Roofs

The home inspector should be able to name the basic style of roof present on the property. There are many styles and almost endless variations of them.

One of the most common roof styles is the **gable roof**. This roof style as shown below is made up of two equal (and opposite) slopes that meet to form a ridge. The term *gable* refers to the triangular areas at each end of the roof. If you'll look back at page 3, you'll see a home with gable roofs that intersect and form a valley. Gable roof homes often have dormers in them with gable roofs of their own.

Take another look at **Photo #1**. This happens to be a steep gable roof. The entryway portion at the center of the house that intersects the rest of the house has a gable roof of its own.

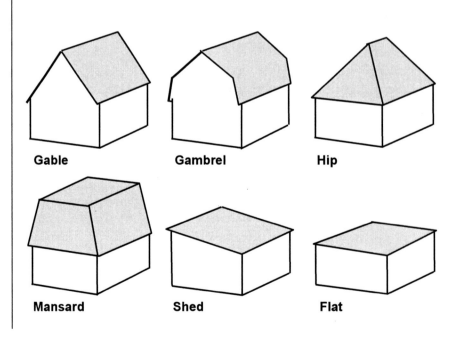

Gable   Gambrel   Hip

Mansard   Shed   Flat

The **gambrel roof** has a double slope on each side, where the bottommost slope is the steepest. This type of roof isn't too common in residential homes, although you'll recognize it as a familiar barn shape.

#2 Gambrel roof

Photo #2 shows an example of a gambrel roof. Notice that the lower slope is much steeper that the upper. We found this roof on a garage. We don't see these roofs on many houses in our area.

The **hip roof** has a slope on all four sides. Each side may meet in a peak at the top of the roof. If a hip roof has a ridge line, then the ridge won't extend the length of the house. Note that the gable roof runs the entire length of the house. A hip roof is named for the hips where adjacent sides meet. A home may have hip roofs that intersect each other.

The **mansard roof** is a variation of the hip roof. They're most often nearly flat on top with steeply sloping sides, although you'll find mansard roofs with the top portion sloped up to a ridge line or to a center peak.

#3 Mansard roof

Photo #3 is an example of a mansard roof. This is not the clearest photo in the world (our apologies), but you may notice that this mansard roof has dormers built into the steep lower slope of the roof. Although it's hard to see, this mansard does not have a flat top. There is a slight upper slope, which you may notice best on the garage.

The **shed roof** is a single slope roof slanting in one direction. You may find a shed roof over an addition (see small addition at the left in the drawing on page 3). Sometimes dormers will have a shed roof instead of a gable.

The **flat roof**, as we said earlier, is most likely not really flat. It's actually a shed roof with a very small slope — less than 2/12.

Natural Enemies

Roof coverings don't last forever. Their natural enemies work against them over time to wear out any roof covering.

- **Sun:** The constant exposure to ultra-violet radiation degrades organic ingredients in roof coverings. The covering heats up and cools down over and over again. Too much sun on wood roofs dehydrates the shingles, causing them to become brittle. Thermal expansion and contraction can destroy adhesion materials in asphalt shingles, for example, and cause cracking in other roofing materials. The southern or southwestern exposures on a home often wear out faster than the northern or eastern.

- **Rain:** Although roof covering systems are designed to protect the roof's structure from water penetration, rain eventually takes a toll on any roof covering. Some ingredients used in roof coverings are soluble and will dissolve over time. Rain washes away granular or gravel finishes in roofing such as asphalt shingles and built-up roofing. Constant wetting of wood shingles can cause them to rot. Metal roofs are susceptible to rust.

- **Wind:** Strong winds can lift shingles off a building. Wind blows rain against a roof and can drive water under the edges of shingles and tiles. Wind can also blow sand against the roof's surface, causing erosion of the covering. With wood shingles, for example, sand erosion can remove enough of the top layer so they no longer protect the shingles underneath.

- **Trees:** We love the cozy effect of trees overhanging the house, but they can do a great deal of damage to a roof. Branches that scrape back and forth over the roof's surface can remove the granules from an asphalt shingle roof. Trees can provide too much shade and, as a consequence, can keep a roof from drying out properly after a rain.

Wood shingles that are not allowed to dry out properly become rotted. Leaves block up drainage systems, causing water damage. Falling branches are an obvious danger to any roof.

- **Moss:** Moss reacts with the organic materials in wood and hastens its breakdown. Wood and built-up roofs are especially vulnerable to the decaying effects of moss. Its root system penetrates the surfaces and creates paths for water to get into and under the roof surface. On other roofing systems, it rusts nails and impedes water runoff.

- **Snow and ice:** A phenomenon called **ice damming** can occur in northern climates when melting snow on the roof refreezes at the roof's overhang. This causes an ice dam to form. Water from melting snow higher up on the roof becomes blocked by the ice dam and cannot escape to the gutters. This water backs up under shingles and seeps into the interior.

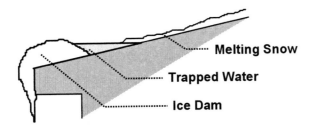

Ice dams occur when enough heat escapes from the attic at the upper part of the roof to melt the snow at the same time the lower part of the roof at the eave is below freezing. Better attic insulation and ventilation would remedy this situation.

Definition

Ice damming is a phenomenon that occurs when melting snow refreezes at the eaves and traps water from snow melting from the upper part of the roof behind it. Ice damming is caused by the temperature differential between the eave, where the temperature is below 32°, and the upper part of the roof, where the temperature is above 32°.

- **Time:** No roof covering lasts forever. We just haven't figured out how to make the ideal roof covering that would never have to be replaced. This chart shows the estimated natural lifetime of various types of roof coverings.

Roof Covering	Life Expectancy
Asphalt shingles	15 to 20 years
Asphalt multi-thickness shingles	20 to 30 years
Asphalt interlocking shingles	15 to 25 years
Roll roofing	10 years
Built-up roofing	10 to 20 years
Wood shingles/shakes	10 to 40+ years
Clay tiles	20+ years
Concrete tiles	20+ years
Slates	30 to 100+ years
Asbestos cement shingles	30 to 75 years
Metal roofing	15 to 40+ years
Single ply membrane	15 to 25 years
Polyurethane with Elastomeric Coating	5 to 10 years

Chapter Three

STRUCTURE FROM THE EXTERIOR

The first step of the exterior roof inspection should be to make a **complete tour around the house**, viewing the roof from all sides. Don't take any shortcuts. Circle the house (and garage) completely and view the roof from all angles. And don't make any assumptions about what the roof looks like at the back of the house or over additions or porches. We've seen houses where the south slope of the roof is in good condition and the north slope completely covered with moss. We've also seen houses where the slope at the front of the house has a new roof covering and the back doesn't.

The home inspector starts the exterior roof inspection by paying attention to the structure of the roof, looking for any exterior signs of structural problems with the roof or the house. This should be done from the ground and from the ladder or the roof's surface. The home inspector should concentrate on the following:

- **The ridge line:** Eye the ridge line(s) on the roof from all angles as you circle the house. A **cracked or broken ridge** can be an indication of structural damage to the roof structure or of structural problems all the way down to the footings. A cracked or broken ridge needs to be further investigated. Sight along the ridge line to see if there is a sag or a hump in it. A **sagging ridge** in an old house may not indicate any other problems. It could just be age or its original design, but it needs to be investigated later from the inside. The sag could be caused by warped rafters pulling the ridge board down. A **hump** shows that the underlying structure is pushing the ridge board up. The home inspector should make note of any problems and search for the cause when the roof's structure is inspected.

- **The planes of the roof:** As you walk around the house, look for any distortions in the planes of the roof. Look at each slope from several angles. Watch for **waviness** across the surface, **sunken areas**, and **raised areas**. Distortions noticed along the roof's surface can be caused by the following problems:

Guide Note

This short chapter presents a discussion of looking for problems with the roof's structure from the exterior. As mentioned earlier, a more detailed presentation of inspecting the roof's structure from the attic is done in another of our guides.

Personal Note

"I came across a house where I noticed a definite slope in the ridge going down toward each side from a center point in the ridge. From the attic, I could see that the ridge board was cracked in the middle.
"The house was actually breaking in half and sinking at each end due to settlement problems."

Roy Newcomer

Definitions

Rafter spread occurs when the roof load bearing on the rafters force them outward.

Truss uplift occurs when the bottom chord of a roof truss bows upward during the cold months and returns to its normal position during the warmer months.

— Wet, rotted, and/or delaminated sheathing
— Sheathing too thin to support the roof covering
— Buckling sheathing that was too closely butted during installation
— Rafters at too wide a span
— Cuts made in rafters or trusses
— Overloaded rafters that are sagging under the weight
— Rafters installed with crowns facing up *and* down
— Warped, cracked, or twisted rafters

The home inspector should make note of any distortions of the planes of the roof and be sure to investigate further from the inside of the attic to find the cause of the problem.

- **The soffits:** The home inspector should look for breaks in the joints between the main house and the roof at the soffits. Breaks or cracks at these joints can indicate:

  — **Rafter spread.** This is a phenomenon where the roof load forces the rafters down and outward. The soffit can be pushed outward along with the rafters, actually separating from the house.
  — **Truss uplift.** The bottom chords of roof trusses can bow and deflect upward during the winter months, pulling ceilings and walls below up with them. In extreme cases, stresses can cause cracking of the soffits as the bottom chords pull them inward.

Make a mental note to investigate any of signs of structural damage when you get into the attic. Also note any signs that it may be dangerous for you to walk on the roof, particularly with distortions that may indicate rotten sheathing.

WORKSHEET

Test yourself on the following questions.
Answers appear on page 18.

1. According to most standards of practice, the home inspector is <u>not</u> required to:

    A. Report methods used to observe roofing.
    B. Inspect the flashings.
    C. Inspect the chimneys.
    D. Inspect attached accessories.

2. Which statement is <u>false</u>?

    A. The customer should not be allowed to get on the roof or ladder.
    B. The inspector shouldn't walk on the roof if it's dangerous.
    C. The inspector doesn't have to walk on the roof if it will damage the roof covering.
    D. The inspector doesn't have to inspect the garage roof.

3. When can a wood roof be safely walked on?

    A. When the roof is dry
    B. During a rain storm
    C. Right after a rain storm
    D. When it's snowing

4. Why should an inspector stay off a tile roof?

    A. A tile roof is too steep.
    B. Tile is too slippery.
    C. Tile is too brittle.
    D. Tile is too spongy.

5. A roof with a slope between 2/12 and 4/12 can be said to be:

    A. Too steep to walk on.
    B. A flat roof.
    C. A conventional roofing system.
    D. A low slope roof.

6. A flat roof has a slope of:

    A. Less than 2/12.
    B. More than 2/12.

7. Identify the roof styles shown here.

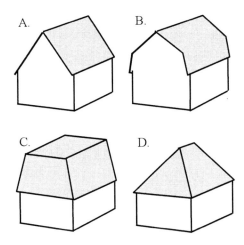

8. What may be an exterior sign of rafters installed at too wide a span?

    A. A break in the joint between the roof and house
    B. A waviness in the roof plane
    C. A sagging ridge line
    D. Truss uplift

9. Rafter spread is:

    A. Where the roof load pushes the rafters outward.
    B. Rafters that are cracked.
    C. A bowing upward of the bottom chord.
    D. Where cuts are made in rafters.

10. Which roof covering is <u>most likely</u> to last the longest?

    A. Asbestos cement shingles
    B. Asphalt shingles
    C. Metal roofing
    D. Roll roofing
    E. Slate shingles

**INSPECTING
ROOF COVERING**

- Approximate age

- Number of layers

- Remaining useful lifetime

- Type of roof covering

- Condition of roof
  covering

Guide Note

*Pages 18 to 43 present
procedures for the inspection of
the roof covering.*

Worksheet Answers (page 17)
1. *D*
2. *D*
3. *A*
4. *C*
5. *D*
6. *A*
7. *A is a gable roof.*
   B is a gambrel roof.
   C is a mansard roof.
   D is a hip roof.
8. *B*
9. *A*
10. *E*

Chapter Four

INSPECTING THE ROOF COVERING

A very important aspect of the exterior inspection is the inspection of the roof covering. The home inspector should **determine the age** of the roof covering. This can be done by first checking with the homeowner as to when the roofing was last replaced. Of course, homeowners aren't always honest, so the inspector should use his or her own judgment to double check. The home inspector should also estimate **how many layers** of roofing are present and estimate its **remaining useful lifetime**. The home inspector identifies the **type** of roofing present and reports on its **condition**. The inspection of the roof covering includes the following:

- Condition of roofing materials
- Condition at ridge, valleys, and interfaces
- Loose or missing components
- Roofing fastenings
- Proper installation
- Water penetration
- Repaired areas

The purpose of the roof covering is to protect the roof framework and the interior from the elements. The roof covering must be able to shed water and to protect the house from the sun.

The home inspector should look over the roof covering while making **an initial tour** around the house, perhaps when looking at the structural aspects of the roof from the ground. The roof covering should be looked at from all sides of the house. The inspector should be noticing the type of roofing, ballparking its age and wear, and making note of areas that look damaged or recently repaired. During this tour, the home inspector may be able to determine whether walking on the roof is possible and where the ladder may be put for access.

The inspector should be creating a **mental plan** of the roof during this tour — various intersecting planes of the roof, dormer areas, lower roofs over additions and porches, and so on. That way the inspector can make sure to inspect all areas once on the ladder at the edge of the roof or on the roof's surface. It is possible to lose track of things you noticed from below once you're on the roof. A mental layout helps you remember.

Asphalt Shingles

Asphalt shingles are made of an asphalt-impregnated felted mat of cellulose or glass fibers which is coated with another asphalt formulation and covered with a granular material. If glass fibers are used, that's called a **fiberglass shingle**.

The asphalt shingle is the most commonly used roof covering today. They come in strips with three tabs or can be individual shingles in various shapes — square butted, hexagonal, or interlocking.

Asphalt shingles used to be graded according to weight. They run anywhere from 210 pounds per square (100 square feet) to 320 pounds per square. The heavier the shingle, the longer life it has. A 225 shingle may last 15 to 20 years, a 320 over 25 years. Today, it's more common to grade these shingles by lifetime, so a manufacturer may offer 20, 25, or 45 year and even lifetime warranty shingles.

Today, most asphalt shingles come with a strip of tar on the surface. The tar lies under the tabs of the next course of shingles. When warmed by the sun, the two courses seal together and are prevented from being blown up by the wind. When roofing is laid in cold weather, the sealing won't take place until the weather warms up again.

Asphalt shingles can be used on a roof with a slope of 4/12 or more using the normal installment techniques. **Felt paper** is normally laid over the roof sheathing as an underlayment. At the eave, a double layer of roofing is laid. The **starter course**, laid under the first course of shingles, is either a solid strip of asphalt composition roofing or a layer with the tabs cut off leaving the tar strip exposed to seal the first course. The shingle strips are laid with the slots alternating. Nails are covered by the course above.

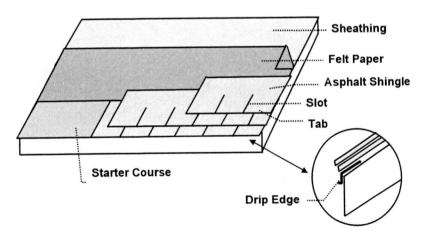

Take the time to visit roofing suppliers and see what types of roof coverings are available. The home inspector must identify the type of covering on the roof with each inspection, and this is a good way to learn how to recognize different types.

INSPECTING ASPHALT SHINGLES

- Loss of granules
- Curling tabs
- Cracking and buckling
- Fishmouthing
- Nail pops
- Loose or missing tabs
- Damaged areas
- Improper installation

An asphalt shingle roof should have a **drip edge** or metal flashing at the lower edge of the roof. This flashing directs the water into the gutters.

A roof with a slope of less than 4/12 (down to 2/12) may use special **low slope asphalt shingles**. When installed, a double layer of felt paper should be used with asphalt cement laid between the courses.

When inspecting the asphalt shingle roof, the home inspector will have to estimate the **number of layers** of shingles. When the roofing wears out, the new shingles may be laid over the old. Local codes put a limit on the number of layers of shingles that can be present. It's commonly only two layers. You can understand why that is. Doubling or tripling the weight per square can put an overload on the roof's structure.

The home inspector should look for the following when inspecting the asphalt shingle roof:

- **Loss of granules:** The granules on the asphalt shingle are what protect the shingle from the ultra-violet rays of the sun. The sun, in fact, is what is responsible for the aging of the shingle. As time goes by, these granules are worn off. As the granules are lost, the shingles dry out and become brittle. Granules are washed away by the rain. The home inspector can look in the gutters and at the bottom of the downspout to see granule deposits to confirm that the roof is losing them.

Photo #4 shows a 15-year-old roof with asphalt shingles. This roofing shows normal wear and tear over time, and it's wearing the way it should be wearing. Some of the corners are starting to come up a bit, but it's still lying fairly flat. Some of the granules are coming off at the top, and it's wearing a bit between the slots. This covering is going to see its 18 to 20 year lifetime. We reported that these shingles would need replacement within the next 5 years.

#4 15-year-old roof with asphalt shingles

- **Curling tabs:** After the granules are lost from the asphalt shingle, water can enter the mat and lead to rapid breakdown. The shingle at the end of its useful lifetime will begin to curl under at the tab edges. This same condition may be caused by poor ventilation in the attic. But basically, when an asphalt shingle roof has curling tabs, it's time to replace them. New shingles cannot be installed over a roof that's gone that far.

#5 Shingles that must be replaced

#6 Shingles with curled tabs

*Photo #5 shows **shingles that must be replaced**. This 15-year-old shingle didn't wear as well as the ones in Photo #4. Notice the curling tabs. Do not walk on shingles like these. You couldn't do it without breaking off every one of those curled corners. The only way you might get up on the roof is to walk next to (but not on) the valleys and ridge line if the shingles there are in better condition. We let the customer know that this roofing needed replacement right away.*

*Photo #6 is another example of **shingles with curled tabs**. Here, you see curling tabs and wide slots that indicate shrinkage. You can even see the tabs underneath. This roof is shot. We should note that this condition isn't necessarily due to age. It can be improper venting. If the homeowner tells you that the roofing was replaced 5 years ago and you see this condition, either the homeowner is lying or the attic ventilation is terrible. We told the customer that new roofing was needed and pointed out that you can't lay new shingles over roofing like this.*

*Photo #7 shows **cracked fiberglass shingles**. Notice how the cracking has taken place down the center of the shingle so that the slots line up with the cracks. Water flowing along the slots will leak right into these cracks. Whenever you see this sort of tear in shingles, chances are that the shingle has a fiberglass mat.*

Personal Note

"Sometimes there's no clue whatsoever that something may be wrong with a roof. One of my inspectors was on a shingle roof that was only 2 years old. Everything looked fine. Suddenly, he went right through the roof and caught himself on the rafters by the armpits.

"The homeowner had patched certain areas of sheathing with 1/4" waferboard, which doesn't meet code. It just can't support a person walking on it.

"I don't mean to scare you. Instead, I'd like you to <u>always be prepared</u> for the unexpected. Don't get too cocky up there."

Roy Newcomer

- **Cracking and buckling:** Fiberglass shingles —shingles with a fiberglass mat — don't respond so well if the roof sheathing moves while expanding and contracting. The normal asphalt shingle can handle the movement, but the lighter fiberglass shingle will sometimes crack. Single shingles or a small area of shingles that are buckled may be the result of warped sheathing.

- **Fishmouthing:** There is a condition called fishmouthing that can occur with asphalt shingles which is caused by excessive overheating, either from multiple layers of shingles or heat buildup in the attic. Shingles can expand so much that they actually separate from the fibers. As a result, the center portion of the tabs curl upwards. As you look up the roof, each shingle looks like a fish's mouth. This condition should not significantly affect the roofing's lifetime.

- **Nail pops:** The nails holding asphalt shingles in place are under the tabs. These nails can pop out. The condition is very easy to recognize. The nails break the seal between courses, pushing up the tabs. You'll see space under the tabs where they should be lying perfectly flat. The nail holes leave paths where water can enter. The condition needs to be fixed.

- **Loose or missing tabs:** You may find a perfectly good roof that is missing one or two strips of shingles. This can happen when the seal between these strips didn't take and they've blown off. Obviously, the strips need to be replaced. The home inspector should test other areas on the roof to see if other shingles are loose.

- **Damaged areas:** Roofing can wear unevenly due to problems with water runoff or the roof's underlying structure. The home inspector should investigate the causes of these problems. Be cautious about stepping on areas like this.

#8 Area of damaged shingles

*Photo #8 shows an **area of damaged shingles**. Here, the problem is one of water runoff from a higher roof to a lower roof. You can see that the homeowner has been making attempts to get the water into the gutter. However, water had been pouring against the two shingles and did extensive damage.*

Photo #9 shows an interesting installation job. See if you can tell what's happened here. First of all, someone's botched this job by lining up the slots which should always be staggered. We don't know if the roofers started at both ends and kept their fingers crossed that they'd be all right when they met in the middle or not. There's been some leaking here. Notice that the slots below the lighter slots are more worn than the others. We lifted the shingles to see what those lighter areas might be. The roofers had put little metal patches in to try to remedy the problem. There was leaking at the patches, by the way. So, this roof was improperly installed. We let the customer know that and informed them that the manufacturer wouldn't warranty such a roof. We suggested that the roof be monitored for leaking and watched carefully. The customer, with luck, might get another 5 years out of it.

Photo #10 shows old T-lock asphalt shingles. We include this as an example of one of the interlocking styles that are available. These shingles are fairly rare and available these days only by special order. They're less prone to leaking, but when the edges where they're locked together wear out, they'll leak. And when they do, it's hard to find the leak.

- **Improper installation:** Shingles should be installed so the slots don't line up, and they should run parallel to the roof line. Nailing must be done properly. The correct length should be exposed to the weather. Any deviations from perfect installation should be reported to the customer.

Here are a few more photos of asphalt shingled roofs. We wanted to include several examples of conditions you might find on this most commonly used roofing material.

#9 Interesting installation job

#10 T-lock asphalt shingles

#11 Moss-covered shingle roof

Photo #11 shows a **moss-covered shingle roof**. The slope of this roof under the trees was completely moss covered. Do not walk on moss-covered areas like this. The part of the roof in the sun may be perfectly safe, but not in the mossy area. It's as slippery as ice. With a roof like this, be sure to check the condition of the sheathing underneath from the attic. Moss can do considerable damage.

#12 Worst shingled roof

Photo #12 is the **worst shingled roof** we've ever seen. The slope at the right is completely deteriorated. This condition is far beyond curling shingles. Here, the shingles are torn, cracked, broken, and have huge pieces just washed away. We wanted you to see this one because the homeowner had re-shingled only half of the roof. Let this one be a lesson to always look at a roof from all sides. Don't make any assumptions about areas of the roof you don't look at.

Personal Note

"Home inspectors sometimes run into people with peculiar ideas. One of my inspectors came across a <u>painted</u> asphalt shingled roof. Sorry, that just doesn't work. The homeowner may have extended the life of the roof by a month or two. That's all."

Roy Newcomer

- Loss of granules

- Buckling and wrinkling

- Lifting laps

- Tearing

- Repaired areas

Definition

The word ply as it applies to roof coverings means a layer of covering. A single ply roof covering has one layer.

Roll Roofing

Roll roofing usually comes in 36" wide rolls and is made of the same material as asphalt shingles — an asphalt impregnated felted or fiberglass mat coated with asphalt formulations and covered with a granular material. Granule coverage varies. Roll roofing is available with complete granular coverage for single ply application or with 50% coverage for double layers.

In **single ply applications** on roofs as low in slope as 2/12, strips are laid parallel to the edge of the roof from eave to ridge with overlapping between laps. Nails are generally left exposed, and seams are sealed with roofing cement. On roofs with a slope between 1/12 and 2/12, a single layer may be used if overlaps are sufficient, nails are concealed, and seams are sealed. These **lower slope roofs** often use two layers, where the bottom ply is completely covered with roofing cement before the top ply is laid.

Roll roofing is an inexpensive roof covering with a short lifetime of only 5 to 10 years. It can be difficult to identify roll roofing since it can be mistaken for a single ply membrane roof (see page 39). Sometimes, roll roofing is used as the cover surface for a built-up roof instead of gravel (see page 27).

The home inspector should inspect the condition of roll roofing for the following:

- **Loss of granules:** As with asphalt shingles, asphalt roll roofing looses its granules over time. Moisture entering the mat causes further breakdown. The surface becomes dull and begins to dissolve.

  Take another look at **Photo #10**. The roll roofing over the porch is dull and is greatly deteriorated at the edge of the roof. The roofing is actually torn. Water had been pouring through into the roof structure of the porch. Not only the roofing but the entire porch had to be replaced.

- **Buckling and wrinkling:** These conditions occur from ongoing expansion and contraction. The buckled and wrinkled areas wear out faster because granules are lost at the wrinkles.

- **Lifting laps:** If you see the edges of the laps or strips of roll roofing lifting up, the seal isn't working. It's likely that water is leaking into these places. Check the roof from underneath to find water penetration through the roof.

- **Tearing:** When roll roofing is torn, it's an indication of movement or warping in the roof sheathing or structure. This condition should be investigated further.

- **Repaired areas:** Homeowners are more likely to try their hand at repairing roll roofing than other kinds. Repairs indicate the homeowner has a problem. Either the strips have been lifting up or there's been leaking into the house. The home inspector should look at repaired areas since homeowners seldom do them right. Repaired areas should be reported to the customer.

#13 Improper repairs

Built-Up Roofing

A flat or slightly sloping roof up to 3/12 can be covered with a covering known as a built-up roof. The built-up roof consists of alternating layers of impregnated **roofing felts and bitumen**. Bitumen is petroleum asphalt or coal tar, although coal tar is seldom used anymore. Each ply is a layer of felt mopped with a layer of asphalt. Built-up roofs can be 2-ply to 5-ply, meaning from 2 to 5 layers of felt and asphalt. The built-up roof is finished with a layer of asphalt formulated for weather resistance. A layer of **slag or gravel** may be applied to the top. (These roofs may be referred to as tar and gravel roofs). Some built-up roofs have a finish layer of roll roofing on them instead of gravel.

***Photo #13** is an example of **improper repairs**. Obviously the homeowner has experienced some problems here. Tar should not be used to reseal this roof. Seams should be repaired with roofing cement placed underneath the laps.*

Definitions

A built-up roof is made of alternating layers of impregnated felt and hot bitumen and topped with a weather resistant coat of bitumen. It may be topped with slag, gravel, or a layer of roll roofing.

Bitumen is petroleum asphalt or coal tar.

Personal Note

"I came across built-up bitumen and gravel roofing on a roof that had more than a 3/12 pitch to it. That's not recommended because under hot conditions the bitumen can run and the plies separate.

"I suggested that a roofing contractor come in to look at it. The contractor evaluated the roof with an infra-red camera to detect areas of heat loss. He could tell where the bitumen was pulling loose from the plywood sheathing."

Roy Newcomer

Photo #14 is an alligatoring built-up roof. This kind of alligatoring and cracking is serious. We advised the customer to get a roofing contractor in there to see how much it would cost to repair or replace the roofing.

The 2-ply built-up roof will last 5 to 10 years; the 4-ply will last 15 to 20. This depends, of course, on periodic maintenance of the surface.

Built-up roofs should be examined for the following problems:

• **Alligatoring:** As the surface of the built-up roof ages and the roofing breaks down, you'll see a network of cracks called alligatoring. The action of the sun will cause the surface to expand and contract until these cracks form. The cracks let moisture into the felts below. A roof with alligatoring needs to be recovered with another layer of asphalt or be replaced.

When the built-up roof leaks, it's very difficult to tell where the leak is coming from. Water can seep into the roofing at one area and run along the plies to emerge through the roof to the interior in a different place.

#14 Alligatoring built-up roof

• **Blistering and bubbling:** When moisture is trapped between the plies through leakage or during installation, blisters or bubbles will form. The plies become delaminated or pull away from each other and from the roof. This reduces the life expectancy of the roofing. Some roofers will cut the blisters open, let them dry out, and reseal them. Others will leave them alone.

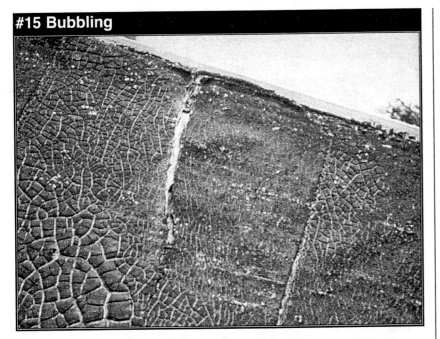

*Photo #15 shows **bubbling** on a built-up roof, not a good sign. We suggested that a roofing contractor take a look at this roof.*

- **Water ponding:** Because built-up roofs are usually flat or gently sloping, water ponding is a problem. The home inspector should report any standing water. If it's dry at the time of the inspection, look for a pattern of dried up ponds left in the dust on the surface.

  Water ponding indicates a structural problem on the flat roof. The roof already has to be sagging under the weight of the roofing for the water to pond there. And the additional weight of the pond will cause more sag. The roof may have to be restructured or proper drains installed to divert the water from this area.

A CAUTION: It isn't always easy to tell just what is going on with a built-up roof. And with the tar and gravel roof, you really can't see what's going on beneath the gravel. When a built-up roof is topped with gravel, the home inspector should report it as *not visible* in his or her inspection report.

Wood Shingles and Shakes

Wood shingles are always sawn, lie flat on one another, and are thinner and more uniform than shakes. **Wood shakes** are thicker with an uneven surface and uneven thickness. Shakes are handsplit, split on one side and sawn on the other, or even sawn on both sides. It gets confusing. Shingles and shakes are usually cedar, although some are made from redwood. They come in lengths of 16", 18", and 24" and vary in thickness. Some are tapered.

Wood shingles are sawn and are of even thickness. Wood shakes are generally thicker with an uneven thickness and rough surface and can be either handsplit, both handsplit and sawn, or sawn.

Heartwood is the wood taken from the inner core of the tree. Edge grain is a cut taken at right angles to the rings of the tree. Clear means that the shingle or shake has no knots or wormholes.

Interlayment refers to the practice of installing roofing felt between each course of shakes. Underlayment refers to a single layer of roofing felt laid over the roof sheathing.

Shingles and shakes are recommended for 4/12 or greater pitched roofs. The greater the slope, the longer they'll last. The lifetime can be up to 50 years, but low grades shingles may last only 15 to 20 years.

Shingles and shakes are sold according to grade. Most codes allow number 1 or number 2 grade only. A quality number 1 grade shingle or shake is cut from **heartwood** (the inner core of the wood), not from sapwood (the outer rings), it's cut from the **edge grain** (at right angles to the rings), and is **clear** without knots and wormholes.

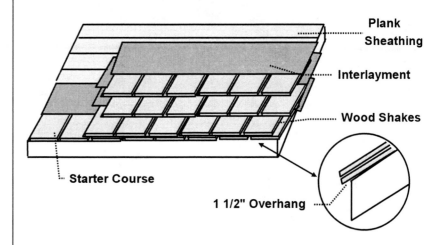

Wood shingles and shakes should be installed over **solid or spaced plank sheathing**. It is not recommended to install them over plywood sheathing. Shakes are laid with a **starter course** under the first course of shakes at the roof's edge with a 1 1/2" overhang as a drip edge. An **interlayment** of roofing felt is laid between each course of shakes. With shingles, the interlayment is not used. Instead, shingles may be laid over an **underlayment** or without the use of roofing felt at all.

Shingles are laid from 1/4" to 3/8" apart; shakes at 3/8" to 5/8" apart. This spacing allows the shingles and shakes to expand when wet. If the proper spacing isn't observed during installation, shingles and shakes can buckle, split, and cup when they expand. Gaps between shingles and shakes should be staggered from course to course. Each shingle and shake is fastened in place with two nails, hidden under the next course. Flashings should be galvanized steel or aluminum; copper is not recommended. We'll talk more about flashings on pages 45 to 54.

When inspecting the wood shingle or wood shake roof, always take the time to determine first whether or not you should walk on the roof. Shingles and shakes that are wet, covered with moss, or mildewed are very slippery. *Do not walk on the roof if any of those conditions are present.* If the shingles or shakes are badly deteriorated, you'll break them if you walk on the roof. Avoid getting on the roof if the condition is bad.

If you do get on the roof, try to walk *across* the roof, not directly up and down from eave to ridge. But be careful, even dry wood roofs in good condition can be tricky.

Start inspecting the wood shingle or shake roof **from the ground**. Looking at the roof from this low vantage point can help you spot areas that are excessively buckled or deteriorated. If the weather is dry, you may notice some curling and shingles or shakes lifted. That's normal. When it rains, they swell up and lay back down.

With the wood roof, the inspection of the roof **from the attic** is very important. The home inspector should be sure to check the type of roof sheathing and determine if it's appropriate. Remember, wood shingles and shakes should have solid or spaced plank sheathing. Spaced sheathing allows the wood to dry out from both sides.

It may surprise the home inspector looking at the roof from underneath if space planking has been used. In dry weather, you may see light coming through from the outside. When it rains and the roofing swells, these spaces will close up. This is fairly common, although the home inspector should look carefully for evidence of water leakage into the attic to be sure the roof isn't really leaking.

During the exterior inspection, the home inspector should inspect the condition of the wood shingle and shake roof for the following:

- **Improper installation:** In dry weather, shingles should not be butted tight against each other and certainly not tight and buckled, split, and cupped. Such shingles are laid without proper spacing. Note that the gaps between shingles and shakes are staggered. Check the overhang at the eaves.

CAUTION

Do not walk on a wood shingle or wood shake roof if the roofing is <u>wet, moss covered, or mildewed</u>. If the shingles or shakes are badly deteriorated, stay off them. They'll break.

For Your Information

There's quite a bit to learn about wood shingle and shake roofs. We recommend that you do some additional reading on the subject. Go to the library or check the Internet. Study up on grading, installation, and chemical treatments applied to wood roofs. If you know a roofer that works with wood roofs, by all means take advantage of the acquaintance to learn more.

INSPECTING WOOD SHINGLES AND SHAKES

- Improper installation
- Softness and rot
- Damaged and weathered
- Loose or missing
- Moss and mildew
- Water penetration

*Photo #16 shows a **deteriorating wood shake roof**. The ends of these shakes are all broken and wearing so thin that the undercourse is exposed. We probed the butt ends and found them soft and rotten. Probing beneath the top course revealed rot underneath. This roof should be replaced, and we advised the customer of that fact. Also, note that this roof is still damp. It's not advisable to get on a roof in this condition. And walking on shakes in this condition can break them.*

- **Softness and rot:** When the wood roof is not allowed to dry out, shingles and shakes can rot. In dry weather, you may see shingles that remain damp. Or you may see those where the butt ends are breaking up, splitting, and cracking. They should be probed for softness and rot at the butt ends and underneath these areas.

- **Damaged and weathered:** Over time, sunlight can dehydrate shingles and shakes, causing them to become brittle and to split and cup. Wind-blown sand can erode the shingle and wear it down. Watch for splits that lie directly under the gap in the course above making a pathway for water to enter the roof. Watch for mechanical

#16 Deteriorating wood shake roof

damage that can occur from rubbing or falling tree branches. You may see evidence of someone else being on the roof. Some roof and gutter cleaners wear spiked shoes or boots on a wood roof for traction. These distinctive holes allow water to get into the shingle. You should wear rubber-soled shoes.

- **Loose or missing:** Check the fastenings on the shingles and shakes. There should be two nails for each one. When nails pull loose, the shingle or shakes may slip out of position. If they're loose, they can be refastened. Keep an eye out for any shingles or shakes that have loosened and blown away. These can be replaced.

- **Moss and mildew:** Moss present on the wood roof should be reported. The most common cause of shingle and shake deterioration is the buildup of moss. If you see moss or algae, be careful, but probe these areas for wood rot. Also check the roof framing below mossy areas, as it can also be damaged from moss. Moss should be removed from the roof, and there are chemical treatments available to kill moss. There are also preservative treatments available that will retard the growth of moss.

  Mildew can also be present. Be careful. Mildew is also slippery. Red cedar normally lightens as it ages. If shingles and shakes are very dark or black, that's a sign of mildew. Mildew can be scraped off but seldom without injuring the shingle itself. There are chemical treatments that can kill mildew.

- **Water penetration:** The home inspector should inspect the wood roof carefully for water penetration. Make note of rotted areas on the exterior and be sure to inspect the roof framing from the attic for any evidence of leaking.

Perhaps the most difficult part of inspecting the wood roof is determining its remaining useful life. A roof in good condition with a long life ahead of it and one that is totally shot are both easy to identify. It's the in-between ones that can be difficult. Start by asking the homeowner how old the roof is and go from there. A 10-year-old roof in bad condition has serious problems and it's aging too fast. Don't estimate 10 more years for it. In general, when about 15% of the roof requires repair, you should recommend a replacement soon.

SOME ADVICE: We suggest that homeowners have a sealant applied to wood roofing — a water resistant stain that includes a mildecide and moss retarder. It prolongs the life of the roof considerably.

AND A CAUTION: There is so much to learn about wood roofs. We suggest that you do further studying on this subject. When commenting on a wood shingle or shake roof, be sure that you're not making *uneducated guesses* about the causes of its condition or suggesting repairs or treatments you're not sure of.

Personal Note

"This is our favorite roofing story. One of my inspectors was up on a flat roof with roll roofing. The house was newly remodeled and everything seemed okay.

"He took a step and his leg went right through the roofing. The good news was that his other foot was firm and kept him from falling through. The bad news was that a swarm of wasps flew out the hole at him, some right up his pants leg.

"Investigating from the attic revealed a 6' by 8' wasp nest. There was <u>no sheathing at all</u> above the nest. (Of course not, the homeowner didn't want to bother the wasps either.) The pest inspector called to deal with the nest froze it, cut it in fourths, and removed it. It now stands on display as the largest wasp nest ever found in our area.

"Our inspector actually finished the inspection, although he said that he felt rather dizzy by the time it was over. He decided to stop at the emergency room on the way home. He had 47 wasp stings!"

Roy Newcomer

Tile Roofs

Tiles are made from concrete or clay and can be flat, curved, or corrugated. Special pieces are made for ridges, valleys, and eaves. Tiles are suitable for a 4/12 sloped roof or steeper. The tile roof may require an underlayment or interlayment.

The tile roof is a heavy, long lasting, fireproof, and weather-resistant roof. It can be up to 4 or 5 times heavier than one with asphalt shingles, and the underlying roof structure must be strong enough to the support the tiles. A tile roof can last from 20 to 50, even 100 years, depending on installation and underlayment.

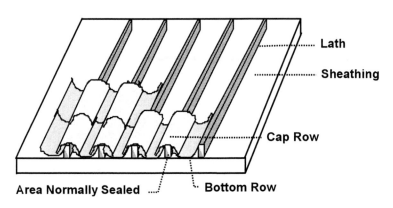

Traditional 2-piece barrel clay tile in the Spanish style are shown above. Support laths are nailed in place over the plank or plywood roof sheathing. A bottom row of inverted tiles is laid between a cap row of tiles. Tiles are overlapped. The fastenings vary. Some tiles are nailed in place, while others have special clips or wire ties. Some tiles are mortared in place. The eaves may be sealed with concrete mortar to keep birds and driven rain out.

*Photo #17 shows a **clay tile roof**. You won't be able to get on a roof like this. Not only is it too steep, but you would break the tiles if you attempted to walk on them. First, view the roof from the ground and see if you can spot any irregularities in the pattern. Tiles can become loose and slip out of place. You may be able to spot broken tiles from the ground. Then, inspect a roof like this from the ladder at the eaves at several places around the house. For example, in this photo, you'd want to get the ladder near the front entrance to get a good look at the chimney area and along any valleys. Clay and concrete tile roofs should be inspected for the following:*

#17 Clay tile roof

The home inspector should not attempt to walk on a tile roof. Walking on the tiles will damage them.

- **Loose, slipped, or missing tiles:** Fasteners can come loose on tile roofing, and tiles can become loose and slip out of place. Slipped tiles are easily noticed and should be reported. They should be repaired. Test other tiles to be sure that fastenings are secure. Any tiles that are missing should be replaced.

  When suggesting repairs or tile replacement to customers, we like to warn them not to call in just any roofer to do the repairs. The homeowner should locate an expert in tile roofs and get an evaluation of the situation before having any repairs done.

- **Broken tiles:** Be sure to point out any tiles that are broken or cracked. Damaged tiles can let water into the roof. Tiles can be broken by mechanical damage. Watch for evidence of tree damage.

- **Prior repairs:** Keep an eye out for previously made repairs. You may see areas where concrete mortar has been applied. If possible, try to determine if the repairs are working. Always report any repairs you see. It's an indication that something has gone wrong. Repairs may not have been done by a reliable or expert roofer. Customers should monitor a roof with previous repairs. It can be a sign that the roofing will need more repairs in the future. Also, they should monitor whether the repair is still working over time.

- **Paint job:** If concrete tiles have been painted (not a common or good thing), inspect the condition of the paint job. Alert customers to the fact that the tiles will need to be repainted every 5 years or so.

Slate and Asbestos Cement Shingles

Slate is sedimentary rock, a natural material, and it should make a long lasting roof. Depending on its quality and thickness (from 3/16" to 2"), slate can last from 30 to over 100 years. One exception is **ribbon slate**, which has a ribbon of color in it. This slate is of lower quality and can crack along the ribbons, lasting only 10 or 15 years. Slate is about 5 times heavier per square than the asphalt shingle roof and must have a strong roof structure to

---

INSPECTING TILE ROOFING
• Loose, slipped, or missing tiles
• Broken tiles
• Prior repairs
• Paint job if present

CAUTION

Do not walk on a <u>slate roof</u> or on <u>asbestos cement shingles</u>. They are brittle and can break.

Definition

A <u>snow shovel</u> is a metal finger-like claw installed with slate roofing to hold snow and ice in place until it melts.

IMPORTANT POINT

If you think you've found slate shingles and they have moss on them, they're really asbestos cement shingles. <u>Moss will not grow on slate</u>.

support it. A proper slope for slate is 6/12 or greater. Slate shingles may have a rectangular exposure or may be laid in diamond patterns.

Slate can be laid on battens over the sheathing or directly on tongue and groove plank sheathing or plywood. Special copper slating nails should be used. In the north, where snow and ice can slide off slate too easily, you'll find **snow shovels**. These are metal finger-like claws that will hold the snow in place until it melts. Snow shovels are installed in a band 1' to 4' wide starting about a foot above the eaves.

Asbestos cement shingles are fairly rare anymore. They're made from a mixture of Portland cement and asbestos fibers and can last from 30 to 75 years. The homeowner needs to be careful during repair or replacement because of the asbestos content in the shingle. The EPA requires that these shingles must be disposed of according to their standards.

Asbestos cement shingles are brittle and may look like slate. These shingles often confuse home inspectors and at times have been improperly identified as slate. Here's one sure way to tell the difference. **Moss cannot grow on slate.** If you think you have a slate roof but it's got moss on it, then you've got asbestos cement shingles.

Both slate and asbestos cement shingles are too brittle to walk on. Don't try to walk on either type of roofing. You'll only damage it. Instead, inspect the roofing from the ladder at several points around the house. The home inspector should inspect the condition of slate and asbestos cement shingles for the following:

- **Loose, slipped, or missing shingles:** Fasteners holding the shingles in place can become loose, causing the shingles to slip out of place. Slipped shingles should be reported so the customer can have them put back in place. If some shingles have slipped, test other shingles for looseness. Report any missing shingles. Check nails, connectors, and clips to be sure they are in place and secure.

- **Broken or flaking shingles:** Because both slate and asbestos cement shingles are brittle, they can be broken or cracked from mechanical damage such as someone walking on them or a falling tree branch. Damaged shingles should be replaced. Low quality slate can **delaminate**, flaking or sloughing off thin layers. When this process continues, the entire exposed portion of the slate can be reduced into tiny bits. A slate roof in that condition would have to be replaced.

Be sure to warn customers to find a qualified and reputable roofer to repair or replace slate. In some states, the roofer may have to be licensed to work with slate.

#18 Asbestos cement shingle roof

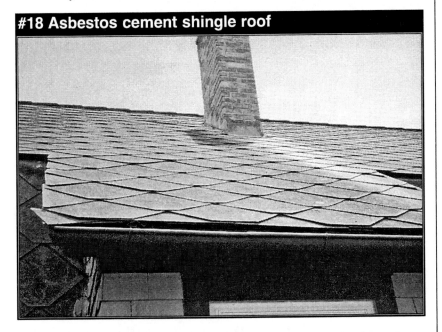

- **Prior repairs:** Watch for repairs. Where repairs have been done, try to determine if the repair was done correctly and if it's working. Report any repairs you find. Customers should be told to monitor a roof with previous repairs.

*Photo #18 shows an **asbestos cement shingle roof**. The loose shingle at the corner was reported. (Be sure to compare this photo to **Photo #19**, which is of a slate roof. Notice how much they look alike.)*

#19 Slate roof

*Photo #19 is a close-up of a **slate roof**. (Compare it to the asbestos cement roof in #18.) There is an improper repair on this roof. The roof was patched, and it was patched wrong. They drilled holes and nailed right through the slate to hold it down. These shingles should not be top-nailed, fasteners should be concealed beneath the slates.*

- Rusting and corrosion
- Secure seams
- Loose sections
- Paint job
- Prior repairs

Definition

Terne metal is a copper containing steel alloy sheet covered with an 80% lead 20% tin plating.

Metal Roofing

Metal roofing may come in sheets, appropriate for flat or sloped roofs, and as shingles for use on sloped roofs only. The metal used may be one of the following:

- Copper
- Stainless steel
- Aluminum
- Galvanized steel in which sheets are covered with zinc
- Coated steel, which is coated with tin, antimony, lead, or nickel alloys
- Terne, which is a copper containing steel alloy sheet covered with an 80% lead 20% tin plating

Metal roofs vary in their expected lifetime of 15 to more than 40 years. But even with a coating, terne and other steel based roofing **will eventually need to be painted** to prevent corrosion and rusting if they are to last their lifetime.

Corrugated galvanized steel is a common metal roofing method. Adjacent sheets are overlapped by one corrugation. Overlapping sheets are laid over the lower course. Each sheet is nailed at the top, not the bottom.

Alloy coated steel sheets are overlapped and can be **soldered** at the seams. Expansion joints must be provided so these seams don't pull apart. Adjacent sheets can also be **folded and crimped** together to form a flat or standing seam. Clips nailed to the roof deck hold the sheets in place. In this case, the lower ends are lapped over and either soldered or waterproofed with roofing cement.

The home inspector may find **metal shingles**. Older homes in the south may have tin plated steel shingles. There are new aluminum shingles, formed and colored to resemble wood shakes, which should be installed over plywood sheathing with a felt underlay.

Because metal roofs are moisture proof, a good ventilation system is needed. The home inspector should carefully inspect the roof framing from the attic for any signs of rot.

The condition of metal roofing should be inspected for the following:

- **Rusting and corrosion:** Inspect for rusting. The lack of paint on steel based and terne roofing is the most likely reason for these roofs to deteriorate. For some reason, it's not common knowledge that they should be painted.

Copper roofing can undergo a chemical reaction to some pollutants in the air, resulting in fine pinholes in the copper, which can cause leaks into the roof.

- **Secure seams:** Inspect metal roofing at the seams, looking for seams that are open, split, or distorted. Check the condition of the solder.

- **Loose sections:** Metal roofing can come loose when seams open or fasteners pop out. Report any loose sections and those that are torn, bent, or punctured.

- **Paint job:** Examine the condition of the paint job on terne and steel based roofing. They should be redone every few years. Be sure to report on the absence of paint on those surfaces that require it.

- **Prior repairs:** People may be tempted to use tar (asphalt roofing compound) to repair open seams or breaks and punctures in a metal roof. But tar should not be used on a metal roof for any reason. The tar traps moisture which only leads to more rusting. The home inspector may even find an entire roof that's been tarred over. This is a very short lived repair job and should be reported as such.

Single Ply Membranes

A recent development in roofing is the single ply membrane, which is made of a modified asphalt base, plastic, or synthetic rubber. These are flexible membranes that cover and seal the entire roof. The membrane may come in strips that are bonded together. They may be precoated with adhesives, laid in

#20 Rubber membrane

For Beginning Inspectors

Travel with a ladder when you visit friends and ask to take a look at the roof. The more experience you have with real roofs, the better off you'll be. We don't have to warn you to be careful, do we?

*Photo #20 shows a shed style roof covered with a **rubber membrane**. This type of roofing is generally used on flat or slightly sloping roofs because it's not an attractive covering. But it is a good roof. Note the little black marks along the edge of the roof. These are the buttons that hold the membrane in place. Notice also the two farthest vents in the roof. These vents should be put in just to vent under the membrane, to get air to the framing so it doesn't rot out. Some types of applications do not require vents.*

adhesives, mechanically fastened with strips or buttons, or held in place with gravel ballast.

We'll talk about flashing more in pages 45 to 54, but for now notice that the area in **Photo #20** where the asphalt shingle meets the siding is tarred. That's not right. There should be step flashing between asphalt shingles and the siding.

The home inspector should inspect the condition of a single ply membrane roof for the following:

- **Poor installation:** Because single ply membranes are new, dating back only to the early 1980s, roofers may not be caught up on the proper installation techniques. Check the membrane to see that it is securely fastened to the roof to prevent liftoff. Seams should be bonded, the surface if glued down should not be overly wrinkled or bubbled, tar should not be used with plastics or rubber, proper venting should be present, and fasteners through the membrane should not leak.

- **Tearing and punctures:** The home inspector should report any tears or punctures in the membrane itself. Because this is a *single ply* roofing material, any break in the membrane will result in leaking into the roof framing.

- **Joints broken:** Sometimes, bonded joints in strip membrane roofs will break open. Again, any opening in the surface will cause leaking. This condition should be repaired.

- **Brittleness:** The single ply membrane is a soft, flexible material, whether of a modified asphalt base, plastic, or rubber. And it should remain flexible. The home inspector should report any evidence that the membrane is becoming brittle.

Reporting Your Findings

As we said on page 6, the #1 rule of home inspection is to take the customer along on the inspection. The #2 rule is to **never take the customer up on the ladder or the roof.** How then does the home inspector manage to communicate his or her findings on the inspection of the roof covering?

Some home inspectors choose to arrive early at the inspection site and begin the roof inspection before the customer arrives. It doesn't hurt to have the customer see you up on the

IMPORTANT POINT
A roof in poor condition is one of the <u>greatest</u> <u>concerns</u> a customer has. Be patient in answering questions and explaining your findings. Be calm when reporting good *and* bad news.

roof hard at work when he or she gets there. That may even be better than asking the customer to stand by and wait while you're up on the roof.

However you manage the inspection of the roof, you still must talk to the customer about its inspection. First, tell the customer why you walked on the roof or inspected it from the ladder. If you didn't get on the roof, explain that the roof was too dangerous or that you would have damaged the roofing if you walked on it. Explain how old you estimate the roof covering to be. Then walk around the house with the customer pointing out the roof from various views and highlighting defective areas as you answer any questions and explain the following:

- **What you were inspecting** — the roof covering (and other aspects of the roof).

- **What you were looking for** — any deterioration of the roof covering, loose fastenings, improper installation, evidence of water penetration, and so on.

- **What you were doing** — probing shingles for rot, checking repaired areas, and so on.

- **What you found** — curled asphalt shingles, wood rot, broken tiles, improper repairs, and so on.

- **Suggestions about dealing with the findings** — replacing missing shingles, applying moss or mildew treatments, repainting a metal roof, and so on. But with this caution — don't make uneducated guesses about how repairs should be made.

- **Estimate of remaining useful lifetime:** With the inspection of the roof covering, customers expect to be told when it will have to be replaced. We suggest that you stick with three general categories — immediate replacement, replacement within 5 years, and more than 5 years of lifetime left. This estimate should be based on the age of the current roof covering *and* your examination.

Be patient with customers when you discuss the roof and be sure to show them defective areas as best you can so they can see for themselves what you're talking about. Be calm and don't blow minor problems out of proportion. A few shingles that need replacing is not a disaster. Help the customer to understand that.

WHEN YOU REPORT

If you're not certain of the roof's age or the number of layers of shingles, use <u>ranges</u> in the written report. For example, use 10 to 15 years and 2 to 3 layers.

Available Reports

The American Home Inspectors Training Institute offers both manual and computerized reports. These reports include an inspection agreement, complete reporting pages, and helpful customer information pages.

If you're interested in purchasing the <u>Home Inspection Report</u>, please contact us at 1-800-441-9411

DON'T EVER MISS

- Missing components
- Loose fastenings
- Wood rot
- Improper installation
- Water penetration
- Repaired areas
- Sag in the ridge or rafters

Filling in Your Report

Every home inspector needs an inspection report. A **written report** is the work product of the home inspection, and every home inspector is expected to deliver one to the customer after the inspection. Inspection reports vary a great deal in the industry with each home inspection company developing its own version. Some are considered to be excellent, while others are not very good at all. A workable and easy to use inspection report is very important for a home inspector in terms of being able to fill it in. Of greater importance is its thoroughness, accuracy, and helpfulness to your customer. We can't tell you what type of inspection report to use, but let's hope that it's a professional one.

The **Don't Ever Miss** list is a reminder of those specific findings you should be sure to include in your inspection report. We list these items after years of experience performing home inspections. Missing them can result in complaint calls and lawsuits later. Here is an overview of what to report in your inspection report regarding the roof inspection:

- **Roof information:** First, identify the style of the roof and the type of roof covering present. Then write the approximate age of the roof and the estimated number of layers of roof present. Also make a note of roof visibility. You may want to give a percentage of visibility or note particular areas of the roof that couldn't be inspected. (We suggest that you report a tar and gravel roof as *not visible*.) It's also a good idea to record where the roof was viewed from — from the roof, the ladder, or from the ground.

- **Roof condition:** Write your findings of the roof covering's condition. We suggest that you use a rating in your report (as mentioned earlier) such as the following:

  — **Satisfactory** indicates the roof covering will last 5 or more years.

  — **Marginal** indicates the roof covering will have to be replaced within the next 5 years. If this is your rating, emphasize this **replacement within the next 5 years** on a summary page of your report as well as on the roof page.

— **Poor** indicates that the roof covering needs to be replaced now or very soon. If this is your finding, you should note this fact as a **major repair** on a summary page of your report as well as on the roof page.

- **Note defects:** Record other findings such as structural defects found in the roof and other problems found in the covering as listed in the Don't Ever Miss list.

GARAGE ROOFING

Don't forget to report the type and condition of the roof covering on the garage too.

WORKSHEET

Test yourself on the following questions.
Answers appear on page 46.

1. Which of the roofs shown in this guide would be okay for the home inspector to walk on?

   A. Photo #6
   B. Photo #9
   C. Photo #11
   D. Photo #16

2. What type of roofing would be appropriate for a roof with a slope of less than 2/12?

   A. Clay tiles
   B. Wood shakes
   C. Single ply membrane
   D. Asphalt shingles

3. What type of roof requires an interlayment of felt between courses?

   A. Wood shakes
   B. Wood shingles
   C. Roll roofing
   D. Slate shingles

4. What is fishmouthing as it refers to asphalt shingles?

   A. Special T-lock shingles
   B. Loss of granules
   C. A special drip edge needed at the eaves
   D. Tabs curling at the center

5. What is a sign that asphalt shingles have reached the end of a useful lifetime?

   A. Tabs curling under at the edges
   B. Nail pops
   C. Granules in the gutters
   D. Loose tabs

6. Moss will grow on slate.

   A. True
   B. False

7. A 4-ply built-up roof would have:

   A. 2 layers of felt and 2 layers of bitumen.
   B. 4 layers of roll roofing.
   C. 2 layers of gravel and 2 layers of asphalt.
   D. 4 layers of felt and 4 layers of bitumen.

8. Which statement about roll roofing is *false*?

   A. It's made of the same material as asphalt shingles.
   B. Problem seams should be tarred over.
   C. It's suitable for flat or low sloped roofs.
   D. Tearing can indicate framing movement.

9. Buckling of wood shingles may indicate:

   A. Rotted shingles
   B. The presence of mildew
   C. Courses not staggered
   D. Improper spacing between shingles

10. Which type of roofing material would most likely be damaged if walked on?

    A. Wood shingles
    B. Asbestos cement shingles
    C. Galvanized steel
    D. Fiberglass shingles

11. How are Spanish style clay tiles installed?

    A. Over spaced planking
    B. Over an adhesive mopped on the sheathing
    C. Over laths nailed to the sheathing

12. What is terne roofing?

    A. Aluminum made to resemble wood shakes
    B. Steel sheets covered with zinc
    C. Copper containing steel alloy sheet covered with lead and tin plating
    D. Copper roofing with pinholes caused by a chemical reaction with pollutants

Chapter Five

FLASHINGS

Guide Note
Pages 45 to 54 present procedures for inspecting the flashings found on the roof.

The roof has a number of vulnerable areas that require more protection than just the roof covering. These areas include:

- At ridges and hips, junctions where planes of the roof slope away from each other
- In valleys, junctions where planes of the roof slope towards each other
- At junctions where a sloping roof meets a flat roof
- At junctions where roof and wall meet
- Around chimneys and other roof penetrations
- At the edges of the roof

Flashings are designed to keep water out of these vulnerable areas. Flashings can be metal — galvanized steel, tin, terne, aluminum, or copper — or roll roofing or the roof covering itself can be used. Flashings must be fastened in place but free to expand and contract.

Ridge and Hip Flashing

Ridge and hip flashings are rarely visible. Metal or asphalt composition is applied to the sheathing across the ridge or hip and the roof covering applied over it. Finishes are then applied over the top. On the asphalt shingle roof, shingles are cut, placed across the ridge, and nailed.

#21 Asphalt shingle ridge cover

*Photo #21 shows an **asphalt shingle ridge cover**.*

Definitions

An <u>open valley</u> is one where the roof covering stops short of the valley and the flashing is visible.

A <u>closed valley</u> is one where the roof covering continues over the valley and the flashing is not visible.

A <u>self-flashed valley</u> is one where the roof covering continues over the valley, but there is no flashing underneath.

Worksheet Answers (page 44)
1. B
2. C
3. A
4. D
5. A
6. B
7. D
8. B
9. D
10. B
11. C
12. C

With the wood shingle or shake roof, special bevel-cut shingles or shakes are used at ridge and hip for a watertight seal. Pieces are laid with alternate overlaps. Nails are concealed. The home inspector should make sure that shingles and shakes in these areas are tight and free of distortion.

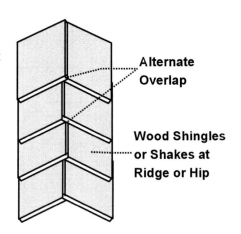

Alternate Overlap

Wood Shingles or Shakes at Ridge or Hip

Slate and asbestos cement shingle roofs may use shingles in the same way as described for wood roofs or have metal flashings capping the ridge. Metal roofs are capped with metal. With roll roofing, a strip of roll roofing is laid across the ridge, glued down with adhesives, and nailed. With built-up roofs, the plies are carried right across the ridge, if there is one. Tile roofs have special cap pieces for ridges, hips, and gables.

The home inspector should check ridges and hips to make sure the material used is watertight. Metal flashings should be inspected for rust and to make sure they are securely fastened.

Valley Flashing

A valley is a trough formed by the junction of two sloping sides of the roof. An **open valley** is one where the roof covering stops short of this trough and the flashing is visible. A **closed valley** is one where the roof covering passes from one face of the roof to the other without break and the flashing underneath is not visible. Shingles at a closed valley may be interwoven or closed cut. A valley where the roof covering passes from one face to another with no flashing underneath is called **self-flashed**. It is not a recommended method.

When **roll roofing** is used as valley flashing, 2 layers are used. An 18" strip is laid facedown and a 36" strip is laid face up on top. Flashings should be cemented to the shingles and nailed to the roof deck.

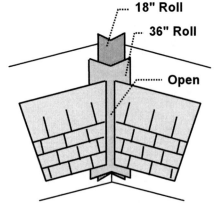

18" Roll

36" Roll

Open

Metal valley flashing is usually 24" wide. It's recommended painted galvanized steel or aluminum be used with wood roofs, not copper. Copper flashings can be found with slate and other roofing materials. Some metal valley flashings are crimped up along the center line to form a rib which reduces the tendency for water to wash from side to side along the valley.

The home inspector may see special valley flashings when water from a steep slope flashing flows into the flashing on a lower slope. These are constructed to keep water from overshooting the lower flashing.

The home inspector should inspect the condition of valley flashings for the following:

- **Poor installation:** Valley flashings should extend to the edge of the roof. If they are too short, water can seep into the eaves. If lengths of flashing are overlapped along the valley, check to see that there is a sufficient overlap. Flashings should lie flat for smooth water flow.

- **Deterioration:** Inspect the flashing materials for corrosion, rust, and abnormal wear. Note if the paint job on painted metals is in good condition.

- **Damaged:** Torn or split valley flashings can be an indication that they are too tightly nailed. Valley flashings can be dented or bent if someone has walked on them. The home inspector should be careful not to damage them either by stepping on them.

- **Repairs:** Note where homeowners have repaired valley flashings indicating a problem area. Try to determine if the repair is working.

  NOTE: Sometimes the home inspector may find valleys that are completely tarred over. Be sure to report tarred valley flashings as *not visible* in your inspection report.

- **Water penetration:** Always check for signs of leaking at valleys. Suspected areas should be noted for checking out once inside the attic.

- **Debris:** Report the presence of debris and leaves in valleys. Suggest that valleys be kept clean so water isn't diverted under the roofing.

INSPECTING VALLEY FLASHINGS
• Poor installation
• Deterioration
• Damaged
• Repairs
• Water penetration
• Debris

*Step flashing, used parallel to
the slope of the roof, is made of
short lengths of overlapped
flashing to form a continuous
sloping flashing.*

*Counter flashing is a second
layer of flashing that covers a
bottom layer of flashing where
roof and wall meet*

Roof-To-Wall Flashing

Flashings are required where second-story walls meet first-story roofs. A flashing is nailed to the wall 6" to 8" up under the siding material and bent to lay over the first 6" to 12" of the roofing material. The home inspector should make sure that the flashing is securely fastened to the wall. This is where leaking commonly occurs.

A **counter flashing** can be used to cover the other flashing. The counter flashing is nailed to the wall under the siding or can be inserted directly into brick siding. See the drawing at the left below.

When the bottom of the wall is on a sloping roof, as at the side of a dormer, then **step flashing** and counter flashing are required. Step flashing consists of short metal flashings that lie under the roof covering and bend up the wall. Step flashings are overlapped at least 3" to form a continuous sloping flashing. A counter flashing is installed over the top of the step flashing and under the siding. See the drawing on the right.

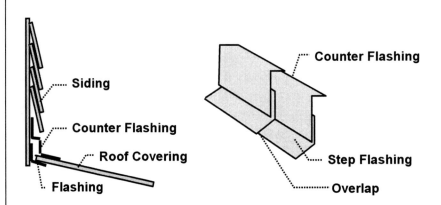

The home inspector should inspect roof-to-wall flashings for the following:

- Poor installation
- Deterioration
- Damage
- Repairs
- Water penetration

Roof-to-Flat-Roof Flashing

Flashings are needed where a sloping roof meets a flat roof. The roof covering of the flat roof — such as roll roofing or built-up roofing — should extend up under the roof covering on the sloped roof. A flashing is then applied at this junction. The home inspector should inspect this flashing for any breaks or tearing, as movement and settlement of the structure can often break these joints.

Joints between a sloped roof and a single ply membrane roof should have metal flashings. The joint should not be tarred. Asphalt is not compatible with plastics and rubber.

Edge Flashing

Flashing at the edge of the roof serves to prevent water from getting into the roof sheathing, backing up behind the roof covering, and seeping into the fascia board. Edge flashing is called **drip edge flashing**. With some roof coverings such as asphalt shingles, drip edges should be present at the eave and at the **rake**. The rake is the overhang at the gable end of the roof. See the drawing on page 19 for an example of a drip edge at the eave.

Edge flashing is nailed to the underlay under the roof covering and extends down to cover the joint between the roof deck and the fascia board. The lower edge is bent out to allow water to drip off into the gutter or away from the fascia board.

#22 Drip edge on a flat roof

Definitions

The rake is the overhang at the gable end of the roof.

A drip edge is a metal flashing used at the eaves and rakes to prevent water from damaging the roof deck and fascia boards. It directs water flow into the gutter.

Photo #22 is an example of a drip edge on a flat roof. The white metal at the edge of the roof is the drip edge. In this case, the roll roofing has been improperly installed. It should go over the top of the drip edge. Notice how much debris is collecting at the junction of the roofing and the flashing. You can bet there is also leaking at this junction.

**INSPECTING
CHIMNEY FLASHINGS**

- Missing or incomplete

- Loose, torn or damaged

- Rusted or corroded

- Prior repairs

- Water penetration

Definition

A <u>cricket</u> is a small peaked roof perpendicular to the main roof slope which is constructed at the high side of the chimney.

Not all roofing systems have drip edge flashing. If it is not present, the home inspector should check at eaves and rakes for water damage, rot, and fungus.

The home inspector should inspect the drip edge for proper installation and function. Check to see that water is not seeping into the roof deck or into the fascia board.

Chimney Flashing

Another vulnerable area on the roof is the junction between the roof and chimney. Flashings are required on all sides of the chimney. They are most often metal, although roll roofing is sometimes used. Strips of roll roofing are laid in adhesives and nailed to the deck and the chimney, often without any counter flashing. The roll roofing can easily pull away from the chimney, and water can get behind the upper edge of the flashing. Metal flashings are preferable.

The **sloping sides** of the chimney will normally have a 2-part system of step flashings and counter flashings as described on page 48. The top of the counter flashing is fitted into the mortar joints on the masonry chimney. The **low side** of the chimney will have typical roof-to-wall flashings as described earlier. The **high side** of the chimney, for chimneys sitting below the ridge, may also have typical roof-to-wall flashing. In this case, flashing should extend up at least 6" or 1/6 the width of the chimney.

When the chimney is wider than 30", a **cricket** should be installed. A cricket is a small peaked roof, perpendicular to the main roof slope and as wide as the chimney, that is constructed on the high side of the chimney to direct water around the chimney. The cricket must have valley flashings on its roof sides and flashings where it meets the chimney.

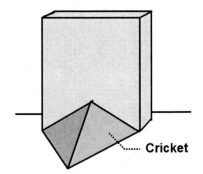

Cricket

The home inspector should inspect the condition of chimney flashings for the following:

- **Missing or incomplete:** Report any missing, incomplete, or inadequate chimney flashings.

#23 Inadequate chimney flashing

*Photo #23 shows **inadequate chimney flashing**. There is no real flashing present at all. They've just put shingles on end against the chimney. This is not acceptable and was reported as such. (Note the use of roll roofing as a ridge cap in this photo.)*

- **Loose, torn, or damaged:** There may be movement of the structure while the chimney stays in place, and flashings can pull away from the chimney.

#24 Loose chimney flashing

*Photo #24 shows **loose chimney flashing**. This flashing is in poor condition. Notice where it has torn and pulled away from the chimney on the high side. Actually, this chimney is greater than 30" wide and should have a cricket behind it. Notice how the debris piles up on the high side of the chimney, which promotes the deterioration of shingles in this area. The side flashing here should be step flashing with counter flashing fitted into the mortar joints. Obviously, they've had problems with leaking here. Notice the patch work.*

- **Rusted or corroded:** Check the condition of the flashing material for rusting and corrosion.

- **Prior repairs:** Be on the lookout for repairs done to chimney flashings. Roofing cement can be used to repair chimney flashing leaks, but it should be considered a temporary solution. Tar is not a proper flashing or repair. Try to determine if the repair was done properly and whether it is working. Suggest that roofing cement patches will need to be monitored and eventually replaced.

#25 Badly repaired roll roofing flashing

*Photo #25 shows **badly repaired roll roofing flashing**. This is really a sorry job. You can't just slop tar all over and expect it to work. We reported that the flashing was in poor condition and should be replaced.*

- **Water penetration:** As with other flashing areas, always check around the chimney for leaks in the flashing.

Penetration Flashing

Areas around piping, vents, fans, and skylights must also have flashing to protect the roof.

- **Pipe and vent flashing:** This type of flashing has a flat part that is laid on the deck and a cylindrical part that extends up and is crimped to the piping. The flat part lies under the roofing on the high side and over the roofing on the low side. Look back at **Photo #21** for several examples of pipe and other types of flashing. These penetrations have been correctly flashed. Tar should not be used around vents and other penetrations.

Aluminum and galvanized steel pipe flashing is available

off the shelf for various pipe sizes and roof slopes. These flashings may also be made of lead or a flexible material. If a flexible flashing is present, the home inspector should feel it to be sure it hasn't become brittle with age.

The home inspector may come across a **pitch pocket**. This is a circular box placed around the plumbing stack, an old method no longer used. The box was filled with a plaster of paris material and covered with bitumen or tar. If you find a pitch pocket, examine it carefully to see if the metal box has corroded, if the "pitch" is not tight to the box, and if the filling has shrunk so much that water is ponding in the box. One or all of these conditions can permit leaking.

- **Ventilators and fans:** These items usually come with flanges that serve as flashings, although the flanges are often not wide enough and may have to be extended or counter flashed. On a sloping roof, the high side of the flange is cemented and nailed under the roof covering, while the lower side lies over it. See **Photo #21** for examples. On a flat roof, the flanges lie under the roof covering on all sides.

- **Skylights:** Many skylights come with a flashing assembly. In some cases, a skylight is a bubble with no flashing (poor quality). The bubble is simply laid under the roofing on the high side and extends over the roofing on the low side. Ideally, skylights should have a curb and be flashed in a manner similar to chimneys. The home inspector should check skylights for leaking, but note that condensation on the interior of the skylight is not necessarily a sign of water penetration.

Roof Structures

Any structure such as dormers, hatchways, bulkheads, and cupolas present on the roof should be properly flashed.

- **Dormers:** Dormer roofs form valleys with the main slope of the roof and should have proper valley flashings as described earlier. The side walls of dormers should have step flashing and counter flashing. The low side of the dormer should have typical roof-to-wall flashing.

ROOF PENETRATIONS
- Piping and vents
- Ventilators and fans
- Skylights

Definition

A pitch pocket is an old method used to flash a plumbing stack on the roof. It is a circular metal box surrounding the plumbing stack, filled with a plaster of paris material and topped with tar or bitumen.

Definition

A parapet is a low wall running above the edge of a flat roof.

Guide Note

The inspection and reporting on roof vents is presented in another of our guides — A Practical Guide to Inspecting Interiors, Insulation, Ventilation.

- **Parapets:** The low wall that runs above the edge of a flat roof is called a parapet. Parapets need flashing at the junction of roof and the wall of the parapet and at the top of the parapet. The flashing at the roof may be typical roof-to-wall flashing or step/counter flashing. The top of the wall should have cap flashing to prevent water penetration. This cap may be asphalt composition, formed metal, brick, or concrete or clay tiles.

  Flashings on the parapet are often in bad condition. They may be loose, deteriorated, or missing completely. The caps should be checked to be sure they're not allowing water to seep into the wall itself.

Reporting Your Findings

After you've finished the roof inspection and are communicating your findings to the customer, be sure to point out areas where you've found problems with the flashing. Show them tarred flashings and valleys and emphasize that you couldn't inspect the flashings underneath the tar. It's important that customers understand what you've found in good condition, what is not in good condition, and what really can't be inspected.

You'll probably be reporting flashings on the same page of your report where you reported roof coverings. Here's an overview of what you should record in your inspection report:

- **Valley flashings:** It's helpful to have a separate area for valley flashings in your report. Identify the type of valley flashings present such as galvanized, aluminum, or asphalt. Then note their condition. Don't report on the condition of valley flashings if you can't see them. (We suggest you report tarred valleys and flashings as *not visible*.)

- **Chimney flashings:** Identify the type of flashing used around the chimney and note its condition.

- **Other flashings:** Again, identify the other types of flashing materials present on the roof and note their condition. Your report may itemize each type of flashing present and allow you to check off satisfactory, marginal, and poor to note the condition of each. In any case, be sure defects, problems, and your recommendations are noted.

WORKSHEET

Test yourself on the following questions.
Answers appear on page 56.

1. What type of ridge cover is used in Photo #21?

   A. Asphalt shingles
   B. Roll roofing
   C. Metal flashings
   D. Drip edge flashing

2. What is an open valley?

   A. One where no flashing is used
   B. One where the roofing covers the valley
   C. One where the flashing is visible
   D. One where the flashing is not visible

3. Torn or split valley flashings may indicate:

   A. They are too tightly nailed.
   B. They need to be crimped at the center line.
   C. They are too short.
   D. They have been repaired.

4. What should the home inspector do upon encountering tarred valleys?

   A. Probe through the tar to see what's underneath.
   B. Report tarred valleys as not visible.
   C. Walk on the tar to see if it's tight.
   D. Report tarred valleys as closed valleys.

5. Where should step flashing be used?

   A. In valleys
   B. At the edge of eave and rake
   C. Around the plumbing stack
   D. Along sloping wall junctions with the roof

6. What is a rake as it relates to roofs?

   A. A tool used to clean debris from the chimney
   B. A flashing at the edge of the roof
   C. The overhang at the gable end of the roof

7. Roll roofing should <u>never</u> be used as a chimney flashing.

   A. True
   B. False

8. Identify the types of flashing indicated in the drawing below.

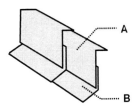

9. What would <u>not</u> be a cause of water backing up under the roof covering at the eave?

   A. Roof covering not extended to roof edge
   B. Valley flashing not extended to roof edge
   C. Missing ridge tiles
   D. A faulty drip edge

10. What is a small peaked roof constructed on the high side of a chimney called?

    A. A pitch pocket
    B. A parapet
    C. A dormer
    D. A cricket

11. A roofing cement repair job on chimney flashing is:

    A. Not recommended.
    B. A temporary solution.
    C. Long lasting.
    D. Absolutely forbidden.

12. Think about this one: *Why is it important to record roof visibility and the methods used to inspect the roof?*

Guide Note
Pages 56 to 65 present
procedures for chimney
inspection.

Chapter Six

INSPECTING THE CHIMNEY

A property being inspected often has more than one chimney. The home inspector reports the **location** of each chimney and then reports on the **condition** of each. The inspector records the **methods used** to inspect the chimney — from the ground with binoculars, from the ladder at the eaves, or from the roof. The inspection of the chimney is conducted not only from the exterior, but from the attic and elsewhere in the house where it is visible. The chimney inspection includes the following:

- Chimney location and clearances
- Chimney foundation
- The material and condition of the chase
- The presence and condition of the chimney cap
- The material and condition of the flue
- Chimney flashings (see pages 50 to 52)
- Condition of the cricket

Chimney Construction

Masonry chimneys are self-supporting structures. Their foundations exist separately from the house foundations and must be deep enough to maintain chimney stability. In cold climates, they should extend below the frost line.

The **chimney chase** is the outer frame of the chimney. It may be constructed of brick, concrete block, stone, metal, or framed with siding. The **chimney flue** is the channel that carries gases, fumes, and smoke from furnaces or fireplaces. It can be made from clay tile, metal, or asbestos cement pipe. The **chimney cap**, usually made of concrete, covers the top of the chase and prevents water and small animals from entering. A **rain cap** may be present over the top of the flue.

Chimneys built before World War II were often **unlined**, meaning that the outer masonry chase itself served as the flue.

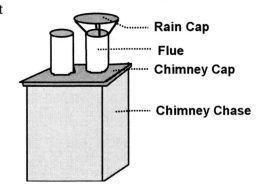

An unlined chimney with its masonry exposed to gases, fumes, and smoke worked well enough then for fireplaces and oil furnaces, but gas furnaces today usually require a separate flue or flue liner, as it is sometimes called.

There are also **metal chimneys**, which are metal from top to bottom. They require special handling such as being isolated from the roof framing, covered with approved fittings, and meeting safety standards on the interior of the house.

Local building codes will generally include regulations regarding the following:

- Height in relation to roof line and neighboring roofs
- Horizontal distance from parapets and other components
- Construction material and thickness
- Type and construction of flues
- Separate flues for each heating system
- Number of flues per chimney
- Construction and depth of chimney footings
- Installation of prefabricated chimneys

Clearances

The home inspector should begin the chimney inspection by making note of location of each chimney present. Chimneys may be located on the slope of the roof, on the ridge itself, or on the end of the roof on the inside or outside of the exterior wall. The location of the chimney must relate to roof lines, both for safety and for the efficient use of the chimneys.

A chimney must extend at least 3' from the point of penetration through the roof. The top of a chimney must be at least 2' higher than anything within 10' of it, including the ridge, dormers, parapets, and so on. If a new addition is built that is higher than an existing chimney, the chimney must be extended to meet these rules. Chimneys not constructed to these standards can experience back-drafting problems.

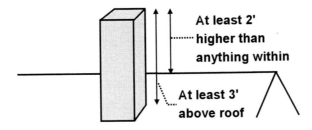

At least 2' higher than anything within

At least 3' above roof

For Your Information

Do some research and find out what your local building codes are regarding chimneys.

INSPECTING CHIMNEY CHASES

- Settling or leaning
- Unlined flue
- Loose, missing, cracked, or spalling masonry
- Deteriorating mortar
- Cracked or missing chimney cap
- Improper or damaged flashing

Inspecting the Chase

The best place to inspect the chimney is from the roof, if possible. It's difficult to know what's happening on the backside of the chimney if you can only see the front. Be sure to record whether you viewed the chimney from the ground, from the ladder, or from the roof. And if you haven't been able to view it from all sides, let your customer know that.

Another point to make is that old, unused chimneys may be only **partially removed**. The chimney may be merely capped off, torn down to below roof level and covered with roofing, or removed entirely. In one case, we found an unused chimney that had been removed *except* for the portion on the roof. The home inspector should determine whether a chimney is no longer in use.

As was stated earlier, the chimney chase may be masonry, made of brick, stone, concrete block, or metal. In newer construction, the chimney chase may be wood framing covered with siding. The home inspector must identify the chase material. Then the chimney should be inspected for the following:

- **Settling or leaning:** The home inspector should inspect the chimney to see if it is settling, pulling away from the house, or leaning. The chimney should be plumb. A chimney sits on its own foundation which can experience the same settlement problems that the main foundation of the house can have. Chimney footings can fail due to soil weakness below. Chimney settlement should be reported to the customer. Deteriorating mortar and masonry can cause a chimney to lean.

- **Unlined chimney:** One of the first things to check on the chimney is if it is unlined — that is, if there is no separate interior flue and the chase itself acts as the flue. This condition should be reported to the customer. In this case, the bricks and mortar of such a chimney are the only barrier preventing the escape of flue gasses.

  CAUTION: If you find an unlined chimney, take great care when inspecting the chase **in the attic**. It's of extreme importance. Any cracked or missing bricks or missing mortar means that carbon monoxide gases are escaping into the attic. This should be reported as a **safety hazard**. When you point out a safety hazard like carbon monoxide escaping into the house, your customer will give a sigh of relief for having hired you. All of a the sudden they've gotten their money's worth.

- **Loose, missing, cracked, or spalling masonry:** Inspect the condition of the chase from the roof *and* from the attic. The chimney takes a lot of abuse and is expected to deteriorate over time. Without the proper protection of a chimney cap, rain constantly wets the interior as well as the exterior. Freeze and thaw cycles, exaggerated by heating cycles, cause masonry to spall. In the absence of a separate flue (unlined), the chase also collects condensation from the gasses as they cool at the top of the chimney.

  Inspect the chase material for any loose or missing masonry pieces, for cracking masonry, and for spalling on the face of the masonry. With metal chases, watch for corrosion and rusting. Inspect framed chases as you would siding.

- **Deteriorating mortar:** The same circumstances that cause the masonry to deteriorate have a detrimental effect on the mortar itself. Mortar should be solid. The home inspector should note any mortar that is loose, broken, missing, or deteriorating.

- **Cracked or missing chimney cap:** A chimney chase with interior and separate flues should have a chimney cap to keep the rain out. A chimney cap prevents moisture deterioration of the masonry and the mortar. The cap should go over the chimney top and fit snugly around the flues. A concrete chimney cap should be at least 2" thick. Sometimes mortar is substituted for concrete, but it will usually crack in a short time. A precast concrete or metal flashing cap may be used to replace a defective cap.

  The home inspector should inspect the chimney cap for cracks. A missing cap should be reported to the customer and a suggestion made to have one installed.

- **Improper or damaged flashing:** Chimney flashing should be inspected. See pages 50 to 52.

Let's study a number of chimney examples. First, look back to **Photo #23**. This chimney has loose and missing mortar. In fact, that's mortar lying on the roof. Notice all the joints where mortar is missing. The bricks are farther apart at the top than at

---

UNLINED CHIMNEY

When you find an unlined chimney, be sure to <u>check the condition of the chase in the attic</u>. Cracks in the chase or missing bricks will permit carbon monoxide to escape into the attic.

Personal Note

"One of my inspectors was up on the roof of an old Victorian that had been extensively remodeled. As he walked past the chimney and rested his hand on it, the chimney swayed slightly. When he turned his back, he heard a horrendous crash. He turned around to find the chimney gone! Only a hole was left. The chimney had fallen 3 stories to the basement.

"When the home was remodeled, the chimney was no longer needed. The homeowners had left the roof portion of the chimney in place while removing the rest of it. However, they hadn't made allowances for supporting the remaining portion on the roof.

"You never quite know what to expect."

Roy Newcomer

Photo #26 shows a *problem chimney*. *Take a look at the chimney cap first. That cap should extend over the top of the top row of bricks, but it only comes to the inside. Also, the chimney cap is cracked and is pulling away from the brick, so leaking has been going on. That's why the bricks along the top are loosening. If fact, a few have already fallen out. What about the height of the chimney?*

This chimney is not 3' higher than the roof at its highest point, and it isn't 2' higher than the roof ridge. It should be. All of these defects were reported.

the bottom of the chimney. At the upper right corner, a brick is moving out of position. Notice also the metal strap around the chimney about 4 rows up from the bottom. This chimney chase is pulling apart. There's also a crack in the chimney cap where rain has been getting in. We reported missing mortar, loose chimney bricks, and a cracked chimney cap as well as improper flashing on this chimney.

Look at **Photo #24** again. This chimney has deteriorating flashing and needs a cricket on the high side. What else is wrong with it? Along the back you can see areas of missing mortar, loose bricks, and spalling on the face of the bricks at the nearest corner.

Looking back, **Photo #25** shows a good close-up of spalling brick. That pile of rubble contains pieces of the face of the brick falling away. There are breaks in the mortar too.

Now let's look at a new photo.

#26 Problem chimney

Photos #27 and #28 are of a **chimney without a chimney cap**.

#27 Chimney without a chimney cap

*Look at **Photo #27**. As we approached this chimney, we could tell that there had been problems. You know how? Notice the white patched mortar at the top. Whenever you see darker mortar at the bottom and tuck pointing at the top, you know the chimney is separating at the top due to mortar deterioration, and it's been patched.*

#28 Close-up of chimney without a chimney cap

Photo #28 is a close-up of the situation inside the chase. Just look how much repair work has been done here. This calls for a mason to evaluate and then repair and replace the flue and cap as necessary.

SOME ADVICE: Don't push or lean on a masonry chimney that's in bad condition. With mortar deteriorating and missing, you're likely to topple it over or knock portions of it down.

INSPECTING THE FLUE

- Requirements

- Cracked, joints open, loose mortar

- Rain cap condition

- Interior soot covered

- Metal chimneys

- Unlined flue

Inspecting the Flue

The flue is the channel that carries gasses, fumes, and smoke from furnaces and fireplaces. When the chimney chase is unlined, then the chase itself serves as the flue. Obviously, the home inspector unable to get on the roof is not going to be able to inspect the chimney flue. The customer should be informed that the flue was *not evaluated*.

If possible, the home inspector identifies the material the flue is made from. The home inspector should inspect the condition of the chimney flue for the following:

- **Requirements:** There should be a 4" separation between multiple flues within the same chase. If not, the flues should be at different heights so as not to interfere with each other's drafts. In general, an unlined chase is not appropriate for a gas furnace, and the home inspector should confirm the presence of an interior flue for a gas furnace.

- **Cracked, joints open, loose mortar:** The flue itself should be inspected for its condition. Check the joints in clay tile flues for loose or missing mortar and cracks in the tiles. In metal flues, joints should be securely closed. A metal flue should be checked for rust and corrosion.

  Look at **Photo #28** again. Here, the junction between the clay tiles is open, and metal strips have been attached to hold the flue together. This flue needs to be evaluated by a mason and repaired or replaced as necessary. The straps are not a proper fix. Once the flue is fixed, a proper method for capping off the chase can be determined.

- **Rain cap condition:** **Photo #27** shows what a rain cap on a flue looks like. The rain cap should be high enough above the flue to allow gasses to escape, about twice the free air space of the flue it is protecting. Inspect the rain cap for condition. It should be firmly attached to the flue and free of cracks, rusting, and corrosion.

- **Interior soot covered:** Most standards do not require the home inspector to inspect the interior of the flue, but we suggest that you do if you can see into the flue. Shine your flashlight into the flue. If it is entirely covered with soot and creosote, report the condition as **not visible**. You should let the customer know that the interior of the flue was **not evaluated** and recommend that a chimney technician be called in to clean the flue and **re-evaluate its condition**.

- **Metal chimneys:** With a metal chimney, where the chimney is also the flue, be sure to inspect it to make sure all sections are tight and securely fastened to each other. Lower sections should be fitted inside upper sections. Watch for rusting and corrosion. Check the metal chimney as it passes through the attic and through the house where possible. Any breaks in the joints will cause the release of carbon monoxide into the house.

  Metal chimneys must be isolated from the roof covering. On the interior, certain clearances must be maintained near combustible surfaces, generally an inch or more.

- **Unlined chimney:** With the unlined chimney, note the interior of the chase for material and mortar deterioration. Anytime you find an unlined chimney, you should recommend that a chimney technician be called to **re-evaluate** it. If you find the interior covered with soot or creosote, report the condition as **not visible**. Again, check the unlined chimney carefully in the attic for cracks, missing masonry, and broken or missing mortar.

The Cricket

If the chimney has a cricket, it should be inspected thoroughly. Check the condition of the roof covering on the cricket. Check valley flashings and flashings against the chimney side. Flashings should be in good condition and water tight.

Other Roof Items

The home inspector will find other items on the roof. Here is an overview of what should be inspected and what is not required to be inspected in the general home inspection.

- **Structures:** Structures on the roof's surface such as dormers and cupolas should be inspected for framing, roof covering, and siding condition just as any other part of the home's structure. Watch for decay, especially with cupolas which are often neglected. Check the condition of the flashings (see pages 45 to 53).

- **Skylights:** Check skylight glazing for cracks, breaks, and discoloration. There should be no condensation between lights. Frames should be free of rot and corrosion. Check for signs of leaking on the interior.

Cupola is a small, rounded structure built on top of a roof; normally used to ventilate attics.

- **Lightning rods, electric service masts, and antennae:** Standards do not require the inspection of these attached accessories. However, the home inspector should check that the lightning rods and cables are securely fastened to the roof. Electric service masts and antennae should sit on pads or protective mountings. Sometimes, the antenna may be strapped to the chimney. Wind vibration can loosen mortar joints in the chimney.

- **Solar systems:** This is not a requirement of a home inspector. But mountings should be tight, and they should be properly flashed and waterproofed. If debris is trapped by the collectors, be sure to examine the condition of the roof covering in this area for deterioration.

Reporting Your Findings

Be sure to talk to your customer about your findings from the chimney inspection. As much as possible, after you come down from the roof, point out those defects from the ground the customer may be able to see. Here's an overview of how you should report the chimney inspection in your inspection report:

- **Chimney information:** First, record where you viewed the chimney, such as from the roof, the ladder, or from the ground. Many homes have more than one chimney. Write the location of each one in the inspection report. You can locate them by describing them, for example, as on the ridge, on the exterior at the north wall, or on the southern slope. If you viewed each chimney from a different place, record that, too.

- **The chimney chase:** Identify the chase material and record its condition, noting defects such as loose bricks, deteriorating mortar, and a cracked chimney cap. Be sure to record your findings of the inspection of the chase from the attic. Note that a chimney in such poor condition that it needs replacement should be noted in your report as a *major repair*.

- **The flue:** If you haven't been able to get up on the roof, mark the flue as *not evaluated*. Identify the flue material and note any defects you've found. Be sure to note if the flue is missing (unlined). We suggest that you always recommend having an unlined chimney cleaned and evaluated by a technician. The same goes for a flue covered with soot and creosote. You should recommend cleaning and evaluation.

- **Safety hazard:** If you find an unlined chimney from which carbon monoxide is escaping into the attic, be sure to list the condition as a safety hazard. It's a good idea to pull out any safety hazards you've mentioned in your inspection report and list them on a summary page at the back of the report. This helps the customer to find them after the inspection.

DON'T EVER MISS

- Loose bricks or stones
- Deteriorating mortar
- Cracked or missing chimney cap
- Unlined chimneys
- Loose or missing flashings
- Cracked flues or open flue joints
- Condition in attic

WORKSHEET

Test yourself on the following questions.
Answers appear on page 68.

1. Identify the items indicated in the chimney drawing shown here.

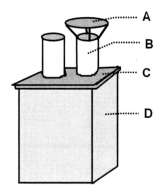

2. In the drawing below, what is the minimum distance indicated by the letter A?

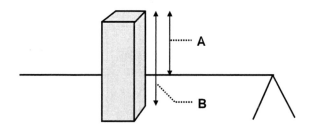

3. In the drawing above, what is the minimum distance indicated by the letter B?

4. What is an unlined chimney?

   A. A chimney chase without a chimney cap
   B. A chimney chase that serves as the flue
   C. A chimney chase with multiple separate flues
   D. A chimney without a chimney chase

5. Which statement is <u>false</u>?

   A. A chimney has its own foundation.
   B. A chimney has its own footings.
   C. A chimney should be plumb.
   D. A leaning chimney must have a footing failure.

6. In Photo #23, what conditions would you report to the customer? *Check as many letters as you wish.*

   A. Missing mortar
   B. Loose chimney bricks
   C. Chimney cap improperly installed
   D. Cracked chimney cap
   E. Missing rain cap
   F. Improper flashing

7. Which photo in this guide would you recommend a cricket?

   A. Photo #23
   B. Photo #24
   C. Photo #25
   D. Photo #28

8. What would <u>not</u> constitute a safety hazard?

   A. A capped off chimney no longer in use
   B. A metal chimney too close to combustibles
   C. An unlined chimney with missing bricks in the attic
   D. An unlined chimney with missing mortar in the attic

9. When should the home inspector recommend that a chimney technician examine the chimney?

   A. When there is spalling masonry
   B. When the interior of the flue could not be evaluated
   C. When flashings are improperly installed
   D. When the chimney cap is cracked

10. The home inspector is required to inspect the condition of lightning rods.

    A. True
    B. False

Chapter Seven

INSPECTING ROOF DRAINAGE

Guide Note

Pages 67 to 72 present the study and inspection of the roof drainage system.

The purpose of the roof drainage system is to allow water to drain off the roof in such as way so as not to damage the house and foundation. Roofs should be pitched in such a way as to allow water to run to the lower edges. Gutter and downspout systems are a common way to divert water away from the roof and house. Flat roofs may have internal drains or scuppers that carry water away.

The home inspector should inspect the following components of the roof drainage system:

- Edge flashings (see page 44)
- Gutters
- Downspouts
- Drains
- Scuppers

Inspecting Gutters

Most homes have gutters attached to the fascia board at the edge of the eaves, some homes have built-in gutter systems, and others may have no gutters at all. Gutters are made from copper, aluminum, galvanized steel, vinyl, or wood lined with metal.

Metal gutters: These gutters may be flat bottomed or half-rounds. They may be hung from brackets, nailed to the fascia board, or spiked to the ends of the rafters.

- **Copper gutters** are the longest lasting — 50 to 100 years — but they are expensive and not that common. Sections of gutter are soldered together. Copper is vulnerable to pitting and deterioration from air pollution and when used with a cedar roof. The home inspector should look for stains on the copper and for broken joints. Orange spots in the metal and rings on the underside will show where water is dripping through.

- **Aluminum gutters** are available in long lengths, so there are fewer joints to worry about. They don't rust but can be dented when ladders are put against them. Special end and corner pieces are used. Joints are usually riveted and sealed. You should inspect them for leaking at these joints.

GUTTER TYPES
• Copper, galvanized steel, and aluminum • Plastic • Wood lined with metal or roofing fabric • Ground-level gutters

Aluminum gutters usually last up to 25 years.

• **Galvanized steel gutters** must be painted to prevent rusting. Some come pre-finished. Joints are usually soldered. Rust is a clue to problems.

Vinyl gutters: Gutters made of vinyl are available. One problem experienced with vinyl gutters is that they become brittle in cold weather. The home inspector should be careful when putting the ladder against them. These gutters vary in expected lifetime because of the quality of the various types available.

Wood gutters: Planked troughs were once used, but now wood gutters are made from clear milled lumber chosen for its straight, tight grain. They can last up to 20 years, although once they begin to decay, their life is limited. Wood gutters may be lined with copper, aluminum, galvanized steel, or lined with roofing compound or roofing fabric, or painted on the inside with boiled linseed oil to fill the pores. They should be carefully probed with a screwdriver for rot, especially at joints and ends where water can enter the end grain. Metal linings deteriorate over time and let water through.

Built-in gutters are constructed as part of the roof. A trough, lined with copper, terne, or galvanized steel, is made at the rafter ends. The metal lining may be replaced with asphalt composition roofing or roofing felt and hot tar. Built-ins can be hard to inspect. There may be leaking even without visible signs of staining. The home inspector should examine them carefully for corrosion of the liner and wood rot.

Yankee gutters, also called flush or Philadelphia gutters, were once very common. They divert water to the edge of the roof and prevent it from running over the eaves. Planks are laid against shaped blocks which are nailed to the roof surface. The fastenings to the roof must be waterproof. The home inspector may find short runs of Yankee gutters over doorways or at the ends of valleys. These gutters may be lined with metal or roofing fabric.

No gutters at all: Some homes were designed without gutters because of the dry climate or in an attempt to solve water runoff without the use of gutters, as was tried in the

Worksheet Answers (page 66)

1. *A is a rain cap.*
   B is a flue.
   C is a chimney cap.
   D is a chimney chase.
2. *2 feet*
3. *3 feet*
4. *B*
5. *D*
6. *A, B, D, and F*
7. *B*
8. *A*
9. *B*
10. *B*

1950's when some homes were designed with extra-wide overhangs. The ground around the house should be examined for a trench showing where the roof water drips off. A determination can be made as to whether the system is depositing water too close to the foundation. Often, the home inspector will find short runs of gutters or diverters that have been added to such homes to solve runoff problems. Don't confuse a home designed to work without gutters with a home where the owner has taken down failed gutters and never replaced them.

Some homes without gutters may have **ground-level gutters**. In this case, a concrete trough in the ground catches the runoff and diverts it away from the house.

The home inspector should inspect the condition of gutters for the following:

- **Corrosion, rust, and rot:** Inspect gutter material and any linings present for deterioration. Check for corrosion, rust, and wood rot at joints and ends where soldering and sealing may break down. Probe wood gutters for wood rot in stained areas and in apparently "good" areas where rot may be present but not visible.

- **Leakage:** Leaks are most likely to occur at joints and at deteriorated areas. Check along the roof edge and fascias for any evidence of leaking and wood rot. Trenches and holes in the ground around the house will show where water is leaking through the gutters. Keep an eye open for repairs to holes and joints and try to determine if those repairs are functioning.

- **Loose fastenings:** Be sure that gutters are not sagging due to loose fastenings. Check to see that brackets are in place and tightly secured. When fastenings loosen up, gutters can change their pitch and may pond water rather than divert it toward the downspout. Water can pour over the edge of the gutter.

- **Damaged:** Gutters can become damaged from trees and ladders leaned against them. Watch for breaks in the gutters and dents or distortions that can divert the flow of water.

- **Debris filled:** Gutters should be cleaned regularly. Debris

INSPECTING GUTTERS

- Corrosion, rust, and rot
- Leakage
- Loose fastenings
- Damaged gutters
- Debris filled
- Wire covers

Personal Note

"I've found gutters so filled with organic matter from leaf breakdown that there are tree saplings growing in them. Of course, the gutters can no longer function. I always stress the importance of keeping gutters cleaned out. Most people don't recognize clogged gutters as a cause of a wet basement."

Roy Newcomer

INSPECTING DOWNSPOUTS

- Corrosion and rust
- Leaking
- Disconnections
- Missing sections
- Clogged with debris

Definitions

The <u>shoe</u> of a downspout is a curved section at its bottom which directs water away from the house

.

A <u>drywell</u> is a rock-filled cavity in the ground near the house. Water is directed from the downspout through piping to a drywell.

and leaves can block the flow of water.

- **Wire covers:** Wire mesh or screens can be installed over gutters to prevent leaves and other debris from entering the gutter. These types of covers do not always perform as they should. Smaller items may still get through the screening and block the gutters. Inspect the covers to see if they are in place and not damaged. Check to see if gutters are blocked anyway.

Inspecting Downspouts

Downspouts connect to the gutters and carry water to the ground or to drains designed for the purpose of disposing of the water. They may be made of the same material as the gutters, although this is not always the case. The bottom end of the downspout may have a shoe and/or a splash block to divert water away from the foundation. Extensions up to 6' in length should be present when the grading is poor around the house.

Some downspouts go **into the ground**. Water may be directed through buried piping to an outlet at some distance from the house. Water could be discharged into the waste disposal system or into basement floor drains. It could be diverted to a **drywell**, which is a rock-filled cavity in the ground near the house. In some cases, piping channels water directly into the ground away from the house. Piping can become clogged with debris and begin to malfunction. The home inspector is not required to inspect downspout underground piping, but it is a good idea to determine which method is being used and whether it is functioning.

The home inspector should inspect the condition of downspouts for the following:

- **Corrosion and rust:** Inspect the condition of the downspout material for deterioration.

- **Leaking:** Look for leaking at the back of the downspout where a broken seam may send water down the siding. Check at connections between sections of the downspout.

- **Disconnections:** It's not unusual to find loose connections between sections of the downspout or to find the shoe disconnected entirely and laying on the ground. When downspouts come apart and are put back together, homeowners often reconnect them incorrectly. The upper section should fit *inside* the lower section.

- **Missing sections:** Often, shoes and extensions to the downspout are missing entirely. Sometimes, vertical sections are missing too, and the downspout simply ends a few feet from the ground. Suggest that any missing sections be replaced.

- **Clogged with debris:** Check both the top and bottom of downspouts for evidence of debris blocking the downspout. Remind customers that downspouts should be cleaned out regularly.

The Flat Roof

A perfectly flat roof will not drain. But in most instances, the so-called flat roof is slightly pitched to allow water to drain off.

In the drawing shown here, the flat roof is constructed with a slight pitch. The flashed parapet along 3 sides of the roof prevents water from escaping. Water is diverted to the roof's edge. A typical gutter and downspout system may be present to catch and carry water away from the roof. If all 4 sides of the roof have parapet walls, **scuppers** will typically be installed through the wall.

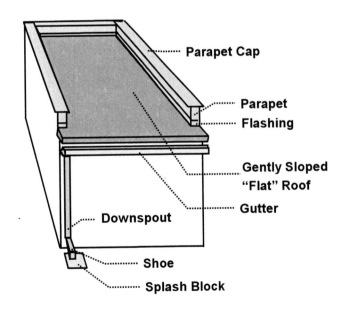

Parapet Cap

Parapet

Flashing

Gently Sloped
"Flat" Roof

Gutter

Downspout

Shoe

Splash Block

When inspecting a flat roof like that shown on page 71, the home inspector should look for evidence of ponding water. Water ponds on the roof when the roof is sagging. If there is no standing water, look for evidence of dried up ponds left in the dust on the surface of the roof.

Flat roofs may be built with reverse slopes to direct the water to a central area for draining. The **butterfly roof** is 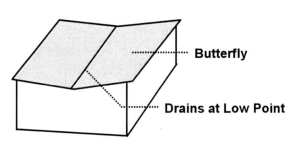 an example. Here, drains at the low intersection of the roof surfaces carry water through interior piping to dispose of or store it.

When **surface drains** are present, the home inspector should try to determine if the drains are working. Cleanout traps, grills, and drain covers should be kept free of debris. Look for evidence around the drain area that water has been ponding around the drain.

Reporting Your Findings

Talk to your customer about what you find when you inspect the roof drainage system. Point out areas in gutters where you find leaking, rusting, or sagging so the customer can see them.

When you report on the gutters and downspouts in your inspection report, identify the materials they're made of. Then report their conditon. Note defects such as those mentioned in the Don't Ever Miss list. If you find gutters clogged with debris, write the suggestion that gutters should be cleaned. Don't overlook reporting on extensions.

The home inspector should also report the condition of gutters and downspouts **on the garage**. Don't forget.

EXAM

A Practical Guide to Inspecting Roofs has covered a lot of detail about types of roof coverings and other roof items such as flashings. Now's the time to test yourself to see how well you've absorbed this information. I included this exam in the guide so you'll have that chance. I hope you'll try it.

To receive Continuing Education Units:
Complete the following exam by filling in the answer sheet found at the end of the exam. Return the answer sheet along with a $50.00 check or credit card information to:

American Home Inspectors Training Institute
N19 W24075 Riverwood Dr., Suite 200
Waukesha, WI 53188

Please indicate on the answer sheet which organization you are seeking CEUs.

It will be necessary to pass the exam with at least a 75% passing grade in order to receive CEUs.

Roy Newcomer

Name_____

Address_____

_____

Phone:_____

e-mail:_____

Credit Card #:_____

Exp Date:_____

Fill in the corresponding box on the answer sheet for each of the following questions.

1. Which action is required by most standards of practice?

   A. Required to walk on the roofing at every inspection.
   B. Required to observe attached accessories such as lightning arrestors and antennae.
   C. Required to inspect underground piping connected to downspouts.
   D. Required to report the methods used to observe the roofing.

2. Which of the following, according to most standards, is not part of the roof inspection?

   A. Flashings
   B. Solar systems
   C. Signs of leaks
   D. Garage roof

3. What is the overall objective of the roofing inspection?

   A. To identify the type of roofing material
   B. To identify major deficiencies in the roofing system
   C. To determine the slope of the roof
   D. To describe the material used as flashings

4. What type of roof has a double slope on each side, where the bottommost slope is the steepest?

   A. Butterfly roof
   B. Gable roof
   C. Gambrel roof
   D. Hip roof

5. Which of a roof's natural enemies can actually degrade organic ingredients in roof coverings?

   A. The sun
   B. The wind
   C. The rain
   D. Trees

6. What is ice damming?

   A. When melting snow refreezes at the eaves trapping water behind it
   B. The use of a metal claw to hold ice until it melts
   C. A wire mesh installed over gutters to prevent ice from entering the gutters
   D. Keeping a roof warm to prohibit ice from forming

7. What does a 3/12 slope mean?

   A. The roof has a 3' run for every 12' of rise
   B. The roof has a 3' rise for every 12' of run

8. Which is the steepest roof?

   A. One with a 2/12 slope
   B. One with a 3/12 slope
   C. One with a 4/12 slope
   D. One with a 6/12 slope

9. Why should an inspector be cautious about walking on a roof over a cathedral ceiling?

   A. They're usually dangerous because of the slope.
   B. The plywood sheathing may be delaminated due to poor ventilation.

10. Assuming the following roofing is in good condition, which could be walked on?

    A. Asbestos cement shingles
    B. Clay tiles
    C. Roll roofing
    D. Slate shingles

11. Which roof covering most likely has the shortest lifetime?

    A. Cement tiles
    B. Metal roofing
    C. Roll roofing
    D. Single ply membrane

12. Identify the roof components marked 1, 2, 3, 4, 5 as shown here:

    A. Dormer, ridge, flue, crown, chase
    B. Valley, ridge, chase, crown, flue
    C. Valley, ridge, flue, crown, chase

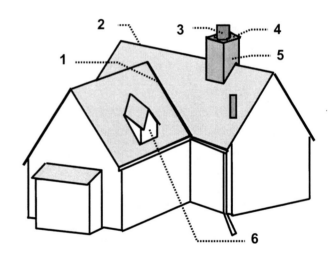

13. In the drawing above, what is the style of the roof on the small addition at the left?

    A. Hip roof
    B. Shed roof
    C. Gable roof
    D. Mansard roof

14. How might rafter spread be noticed from the exterior of the home?

   A. A hump in the ridge line
   B. Waviness over the roof surface
   C. Separation of the soffits from the wall
   D. A sunken area on the roof

15. What is truss uplift?

   A. Where the roof load pushes the rafters upward
   B. Trusses at too wide a span
   C. Trusses that have had chords cut
   D. A bowing upward of the truss's bottom chord

16. When inspecting roof structure from the exterior, the home inspector should:

   A. Make a complete tour around the house to view the roof from the ground.
   B. Eye the ridge lines from the ground.
   C. Look for distortions in the planes of the roof.
   D. All of the above.

17. What roof condition might be part of the original design of the roof?

   A. A hump in the ridge line
   B. A sagging ridge line
   C. A sunken area in the roof surface
   D. Waviness in the roof surface

18. Which of the roofs shown in the guide would pose a danger to the inspector who tried to walk on it?

   A. Photos 1, 11, 18
   B. Photos 1, 11, 14
   C. Photos 11, 14, 18
   D. Photos 5, 11, 18

19. What does 225lb asphalt shingle roofing mean?

   A. It costs $2.25 per strip
   B. It weights 225 pounds per 100 square feet
   C. 225 shingles cover 100 square feet
   D. It lasts 22.5 years

20. What type of roofing would not be appropriate for a roof with a slope of 5/12?

   A. Built-up roofing
   B. Wood shakes
   C. Cement tiles
   D. Asphalt shingles

21. Reporting the condition of the roof covering to be marginal means:

   A. The roof should be replaced now.
   B. The roof should be repaired now.
   C. The roof should be replaced within 5 years.
   D. The roof will reach its normal lifetime.

22. The home inspector is NOT required to report:

   A. Approximate age of the roof covering
   B. Number of layers of the roof covering
   C. The cost of replacing the roof covering
   D. All of the above

23. Finding granules from asphalt shingles in the gutters definitely indicates:

   A. Roofing has reached its useful lifetime.
   B. Roofing was improperly installed.
   C. Roofing has experienced excessive overheating.
   D. Roofing is showing some wear.

24. Roll roofing has the same materials as:
   A. Asbestos cement shingles
   B. Concrete tile
   C. Asphalt shingles
   D. Terne roofing

25. Which statement about wood shingles is false?

   A. Wood shingles should be installed over solid or spaced planked sheathing.
   B. Wood shingles should be laid 1/4" to 3/8" apart.
   C. Wood shingles should be staggered.
   D. Wood shingles should be installed with an interlayment of felt between each course.

26. If there are splits in wood shakes directly under the gaps in the course above, the home inspector should look for:

   A. Loose nails at the splits.
   B. Softness and rot at the splits.
   C. Shakes butted too closely at the splits.
   D. Missing interlayment.

27. Darkened or blackened wood shingles is an indication of:

   A. Wood rot
   B. Moss
   C. Mildew
   D. Water penetration

28. If recommending repairs to a tile roof, the home inspector should warn the customer:

   A. To find an expert for the job.
   B. The repairs will be expensive.
   C. That repairs hardly ever work.

29. What is a sign of low quality slates?

   A. The presence of moss on them
   B. Brittleness
   C. Slipped shingles
   D. A ribbon of color in the slate

30. Slate and asphalt shingles look very much alike.

   A. True
   B. False

31. Which metal roofing does not need to be painted?

   A. Terne
   B. Copper
   C. Coated steel
   D. Galvanized steel

32. The inspector finding tar used on a single ply rubber membrane roof should:

   A. Check to see if the tar seal is working.
   B. Check to see if the membrane is still securely fastened to the roof.
   C. Tell the customer tar should not be used with rubber roofing.
   D. Recommend the inspector come back to remove the tar.

33. Which statement is true?

   A. Copper valley flashings should be used with wood shingles.
   B. Galvanized steel valley flashings should be painted.
   C. Roll roofing is not suitable for use as flashing.
   D. An open valley means no flashing is present.

34. What is a counter flashing?

   A. A second layer of flashing that covers a bottom layer of flashing
   B. A bottom layer of flashing covered by step flashing
   C. The kind of flashing used at the rake edge of the roof
   D. A special ridge cover used with tile roofs

35. In Photo #22, what condition should be reported?

   A. Drip edge flashing is missing.
   B. Roll roofing is not proper for this roof.
   C. Roll roofing should cover the drip edge.

36. What is the item noted in the chimney drawing below?

    A. Parapet
    B. Chimney chase
    C. Cricket
    D. Dormer

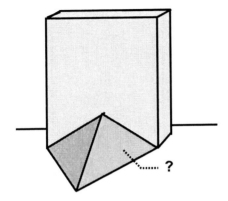

37. What defects should be reported for the chimney shown in Photo #26 in the photo guide?

    A. Chimney cap should be extended over bricks, chimney cap is cracked, step flashing should not be used here
    B. Chimney cap is cracked, loose and missing brick, step flashing should not be used here
    C. Chimney cap should be extended over bricks, chimney cap is cracked, there are loose and missing bricks
    D. There are loose and missing bricks, chimney cap is cracked, step flashing should not be used

38. What does it mean if a chimney is unlined?

    A. There is no chimney cap covering the top of the chimney chase.
    B. There is no chimney chase present.
    C. The chimney chase itself serves as the flue.
    D. The chimney is made of metal and doesn't need a flue.

39. In the attic area, cracks or missing bricks in the unlined flue should be reported as:

    A. Not visible.
    B. A safety hazard.
    C. Not evaluated.
    D. Needs cleaning.

40. According to the standards of practice, the home inspector is required to inspect attached accessories for:

    A. Proper installation.
    B. Their effect on the roofing.
    C. Their condition.
    D. Proper functioning.

41. Which gutters, if properly maintained, can be expected to last the longest?

    A. Copper
    B. Aluminum
    C. Plastic
    D. Wood

42. Identify the types of roofs marked 1, 2, 3. 4 as shown here:

    A. Gable, hip, mansard, shed
    B. Hip, gable, mansard, butterfly
    C. Hip, gable, mansard, shed

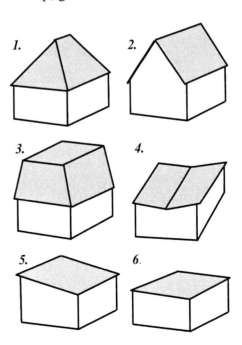

43. Given the type of roof shown in the drawing below, what elements contribute to protecting against water penetration?

   A. Splash block, parapet cap, slight slope to roof
   B. Tight connection in the downspouts, parapet cap, splash block
   C. Splash block, parapet cap, roofing in good condition
   D. Parapet cap, flashing between parapet wall and roof, roofing in good condition

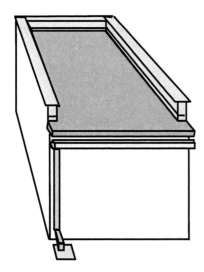

44. Which statement is <u>false</u>?

   A. Some downspouts may go into the ground.
   B. Some houses may have been designed to work without gutters.
   C. Galvanized steel gutters should be painted
   D. Aluminum gutters should be inspected for rust.

45. What is a drywell?

   A. A well that is no longer used
   B. A underground cavity that receives water runoff from the roof
   C. A concrete trough in the ground around the house that receives water runoff from the roof
   D. An unused underground oil tank.

46. **Case study 1:** You are inspecting an asphalt shingle roof in the winter. Although the north slope of the roof is snow covered, the south side is not. The roofing is 15 years old but shows only normal wear and tear. Valley flashings are tarred.

   Do you walk on the south side of the roof?

   A. Yes
   B. No

47. For case study 1, what do you report roof visibility in your inspection report.

   A. All
   B. None
   C. Percent 50% Limited by snow

48. For case study 1, how would you report the condition of the roof?

   A. Satisfactory
   B. Marginal
   C. Poor

49. **Case Study 2:** You are inspecting a masonry chimney which is unlined. Masonry and mortar are in good shape on the exterior but the inside is soot covered. In the attic, you notice missing bricks and mortar in the chimney.

   How would you report the flue in the inspection report?

   A. Unlined, have cleaned and re-evaluated
   B. Not evaluated, unlined
   C. Flue in good condition, not evaluated
   D. Unlined, in good condition

50. For case study 2, what, if any, condition would you report as a safety hazard?

   A. Missing bricks in attic area of chimney
   B. A soot covered flue
   C. Unlined flue

GLOSSARY

Alligatoring A condition with built-up roofing where a network of cracks covers the surface due to expansion and contraction.

Asbestos cement shingles A mixture of portland cement and asbestos fibers in shingle form, used as a roof covering.

Asphalt shingles Asphalt-impregnated felted mats coated with an asphalt formation and covered with a granular material, used in shingle form as a roof covering.

Bitumen Petroleum asphalt or coal tar.

Built-in gutters A trough built as part of the roof at the rafter ends, normally made of wood lined with metal.

Built-up roofing Alternating layers of impregnated felt and hot bitumen topped with a weather resistant coat of bitumen, used as a roof covering. Usually topped with gravel, slag, or a layer of roll roofing.

Butterfly roof A roof style with two opposite slopes sloping inward toward the center of the roof.

Chase See *Chimney chase.*

Chimney cap A concrete covering for the top of the chimney chase.

Chimney chase The outer construction that encloses the flue.

Chimney flashing Metal or roll roofing flashings used in the joint between chimney and roof to prevent water penetration.

Chimney flue The channel that carries gases, fumes, and smoke from heating units and fireplaces.

Clear In wood shingles or shakes, wood with no knots or wormholes.

Closed valley A valley where the roof covering continues over the valley and the flashing is not visible.

Coated steel Steel sheeting coated with tin, antimony, lead, or nickel alloys, used as a roof covering.

Counter flashing A second layer of flashing that covers a bottom layer of flashing where roof and wall meet.

Course Each row of roofing material.

Cricket A small peaked roof perpendicular to the main roof slope constructed at the high side of a chimney.

Dormer A structure built out from the roof slope having its own roof and walls.

Downspout The vertical piping connected to gutters that carries rainwater from the roof to the ground.

Drip edge A flashing laid at the eaves and rakes to direct water flow into the gutters or off the roof.

Drywell A buried gravel pit that accumulates water and allows it to seep into the ground slowly.

Edge grain A cut of wood taken at right angles to the rings of a tree.

Felt paper An underlayment laid on roof sheathing under the roof covering.

Fiberglass shingles Mats of glass fibers covered with a granular material, used in shingle form as a roof covering.

Fishmouthing A condition with asphalt shingles where the center portion of the tabs curl upwards due to overheating.

Flashings Sheet metal used at joints between building components to prevent water penetration.

Flat roof A variation of a shed roof with a very small slope, less than 2/12.

Flue See *Chimney flue.*

Gable roof A roof style made up of two equal and opposite slopes that meet to form a ridge.

Galvanized steel Steel sheeting covered with zinc, used as a roof covering.

Gambrel roof A roof style with a double slope on each side where the bottommost slope is the steepest.

Ground-level gutter A concrete trough in the ground that catches rainwater runoff from the roof and carries it away from the house.

Gutter A metal, plastic, or wood trough attached to the roof at the eaves to collect rainwater and carry it away from the roof.

Heartwood The wood taken from the inner core of a tree.

Hip roof A roof style with a slope on all four sides, either meeting in a peak or a ridge line.

Ice damning A phenomenon that occurs when melting snow refreezes at the eaves and traps water from snow melting from the upper part of the roof.

Interlayment Roofing felt installed between courses of wood shakes.

Laps Strips of roll roofing.

Mansard roof A variation on the hip roof with steeply sloping sides and a top portion either flat or meeting in a peak or a ridge.

Open valley A valley where the roof covering stops short of the valley and the flashing is visible.

Parapet A low wall running along the edge of a flat roof.

Pitch pocket A circular metal box surrounding the plumbing stack, filled with a plaster of paris material, and topped with tar or bitumen, used as stack flashing.

Ply A layer of roof covering.

Rafter spread A phenomenon where the roof load bearing on the rafters force them outward.

Rain cap A metal cap attached to the top of the chimney flue allowing gasses to escape but prevents rain from entering the flue.

Rake The overhang at the gable end of the roof.

Ribbon slate Slate shingles with a ribbon of color in them.

Ridge The horizontal intersection of two sloping roof surfaces.

Ridge cover Protective flashing laid across the ridge and finished with shingles or roll roofing to protect the ridge from water penetration.

Rise The vertical height of the roof.

Roll roofing An asphalt-impregnated felted mat of fibers coated with asphalt formations and covered with a granular material, used in strips as a roof covering.

Roof covering The outer layer of shingles, tiles, or other materials used to protect the roof from water penetration.

Run The horizontal length of the roof from the eave to the center point.

Self-flashed valley A valley where the roof covering continues over the valley, but there is no flashing underneath.

Sheathing Sheets of plywood or wood planking used to cover the roof frame.

Shed roof A roof style with a single slope slanting in one direction.

Shoe The curved section at the bottom of the downspout which directs water away from the house.

Single ply One layer of roof covering.

Single ply membrane A modified asphalt, plastic, or rubber membrane laid in adhesives or mechanically fastened, used as a roof covering.

Slate shingles Sedimentary rock in shingle form, used as a roof covering.

Slope The ratio of a roof's rise to its run. Normally expressed with the measure of the rise over a run of 12'. A slope of 3/12 is said, "Three in twelve."

Snow shovel In roofing, a metal finger-like claw installed with slate roofing to hold snow and ice in place until it melts.

Soffit The horizontal underside of the eave.

Square The amount of roofing material used to cover 100 square feet of roof surface.

Starter course Shingles or shakes laid under the first course of shingles at the edge of the roof.

Step flashing Short lengths of overlapped flashing to form a continuous slopping flashing, used parallel to the slope of the roof.

Tab In a shingle roof, each shingle.

Terne metal A copper containing steel alloy sheet covered with an 80% lead, 20% tin plating, used as a roof covering.

Tile Concrete or clay flat, curved, or corrugated shaped forms, used as a roof covering.

T-lock shingles A rare style of interlocking asphalt shingles.

Truss uplift A phenomenon where the bottom chord of a roof truss bows upward during the cold months and returns to its normal position during the warmer months.

Underlayment Roofing felt laid in a single layer between the roof sheathing and the roof covering.

Unlined flue A chimney where the chase itself serves as the flue.

Valley The trough formed by the junction of two sloping sides of the roof.

Valley flashing Metal or roll roofing flashing laid in a valley to protect the junction from water penetration.

Wood shakes Thick, rough, uneven shingles that are handsplit, split and sawn on one side, or sawn on both sides, used as a roof covering.

Wood shingles Shingles that are sawn and are of uniform thickness, used as a roof covering.

Yankee gutters Planks nailed to the roof surface to prevent rainwater from escaping from the roof in areas such as over doorways.

INDEX

A Practical Guide to Inspecting Program
Study Unit Three, Inspecting Roofs

Student Name: _____ Date: _____

Address: _____

Phone: _____ Email: _____

Organization obtaining CEUs for: _____ Credit Card Info: _____

After you have completed the exam, mail *this exam answer page* to American Home Inspectors Training Institute. You may also fax in your answer sheet. You will be notified of your exam results.

Fill in the box(es) for the correct answer for each of the following questions:

1. A☐ B☐ C☐ D☐	24. A☐ B☐ C☐ D☐	47. A☐ B☐ C☐
2. A☐ B☐ C☐ D☐	25. A☐ B☐ C☐ D☐	48. A☐ B☐ C☐
3. A☐ B☐ C☐ D☐	26. A☐ B☐ C☐ D☐	49. A☐ B☐ C☐ D☐
4. A☐ B☐ C☐ D☐	27. A☐ B☐ C☐ D☐	50. A☐ B☐ C☐
5. A☐ B☐ C☐ D☐	28. A☐ B☐ C☐	
6. A☐ B☐ C☐ D☐	29. A☐ B☐ C☐ D☐	
7. A☐ B☐	30. A☐ B☐	
8. A☐ B☐ C☐ D☐	31. A☐ B☐ C☐ D☐	
9. A☐ B☐	32. A☐ B☐ C☐ D☐	
10. A☐ B☐ C☐ D☐	33. A☐ B☐ C☐ D☐	
11. A☐ B☐ C☐ D☐	34. A☐ B☐ C☐ D☐	
12. A☐ B☐ C☐	35. A☐ B☐ C☐	
13. A☐ B☐ C☐ D☐	36. A☐ B☐ C☐ D☐	
14. A☐ B☐ C☐ D☐	37. A☐ B☐ C☐ D☐	
15. A☐ B☐ C☐ D☐	38. A☐ B☐ C☐ D☐	
16. A☐ B☐ C☐ D☐	39. A☐ B☐ C☐ D☐	
17. A☐ B☐ C☐ D☐	40. A☐ B☐ C☐ D☐	
18. A☐ B☐ C☐ D☐	41. A☐ B☐ C☐ D☐	
19. A☐ B☐ C☐ D☐	42. A☐ B☐ C☐	
20. A☐ B☐ C☐ D☐	43. A☐ B☐ C☐ D☐	
21. A☐ B☐ C☐ D☐	44. A☐ B☐ C☐ D☐	
22. A☐ B☐ C☐ D☐	45. A☐ B☐ C☐ D☐	
23. A☐ B☐ C☐ D☐	46. A☐ B☐	